Advanced Intelligent Environments

Wolfgang Minker · Michael Weber · Hani Hagras ·
Victor Callagan · Achilles D. Kameas
Editors

Advanced Intelligent Environments

Springer

Editors
Wolfgang Minker
University of Ulm
Institute of Information Technology
Albert-Einstein-Allee 43
89081 Ulm
Germany
wolfgang.minker@uni-ulm.de

Michael Weber
University of Ulm
Institute of Media Informatics
Albert-Einstein-Allee 11
89081 Ulm
Germany
michael.weber@uni-ulm.de

Hani Hagras
Department of Computer Science
University of Essex
Wivenhoe Park
Colchester
United Kingdom CO4 3SQ
hani@essex.ac.uk

Victor Callagan
Department of Computer Science
University of Essex
Wivenhoe Park
Colchester
United Kingdom CO4 3SQ
vic@essex.ac.uk

Achilles D. Kameas
Hellenic Open University & Computer Technology Institute
N. Kazantzaki Str.
265 00 Patras
University Campus
Greece
ie2006@cti.gr; kameas@eap.gr

ISBN 978-0-387-76484-9 e-ISBN 978-0-387-76485-6
DOI 10.1007/978-0-387-76485-6
Springer Dordrecht Heidelberg London New York

Library of Congress Control Number: 2008944170

Printed on acid-free paper

Springer is part of Springer Science+Business Media (www.springer.com)

Contents

Contributing Authors

Yacine Bellik is an assistant professor at the University of Paris-Sud, France. He leads the "Multimodal Interaction" research topic at LIMSI-CNRS (Laboratoire d'Informatique pour la Mécanique et les Sciences de l'Ingénieur, France). He holds a Ph.D. and Habilitation à Diriger des Recherches (HDR) in computer science. His research interests concern multimodal human–computer interaction, aid for the blind and ambient intelligence.

Raja Bose is a member of research staff at Nokia Research Center Palo Alto, USA. He received his ph.D. in Computer Engineering from the Mobile and Pervasive Computing Laboratory at University of Florida, USA. He is currently engaged in research activities involving mobile device interoperability in intelligent environments. His other research interests include sensor networks and application of service-oriented architectures and complex event processing in smart spaces. He is a member of IEEE.

Yolaine Bourda is a professor in computer science at Supélec (École Supérieure d'Électricité), France and Head of the Computer Science Department. Her research interests include adaptation to the user. She is a member of the ISO/IEC JTC1/SC36 committee and co-editor of the standard ISO/IEC JTC1 19788-1 (Metadata for Learning Resources).

Victor Callaghan is a professor of computer science at the University of Essex, UK, where he leads the Inhabited Intelligent Environments Group and is a director of the Centre for Digital Lifestyles. Professor Callaghan was responsible for setting up the University's Brooker Laboratory (for embedded systems), Robot Arena (for mobile robotics) and Space (a digital home testbed), topics which form his research focus and which he has published over 100 papers on. He is a member of the editorial board of the *International Journal of Pervasive Computing and Communications* (JPCC), associate editor of the *International Journal of Ubiquitous Computing and Intelligence* (JUCI) and a member of the editorial board of the *Intelligent Buildings International Journal* (IBIJ). He has served as programme chair of Pervasive Computing and Applications 06, co-chair of Ubiquitous Intelligence and Smart Worlds 05 and 06 and Intelligent Environments 05, 06, 07 and 08.

Carl K. Chang is a professor and chair of computer science at the Iowa State University, USA. He received his Ph.D. in computer science from Northwestern University in Evanston, Illinois, USA, in 1982. His research areas include software engineering, services science, and net-centric computing. He worked for Bell Laboratories from 1982 to 1984, and joined the University of Illinois at Chicago, USA, in 1984 to begin his academic career. He served as the editor-in-chief for *IEEE Software* (1991–1994), was elected the 2004 president of the IEEE Computer Society, and is now serving as the editor-in-chief for *IEEE Computer*.

Diane J. Cook is currently a Huie-Rogers Chair Professor in the School of Electrical Engineering and Computer Science at Washington State University, USA. She received a M.S. degree from the University of Illinois, USA, in 1987 and a Ph.D. degree from the University of Illinois, USA, in 1990. Dr. Cook currently serves as the editor-in-chief for the *IEEE Transactions on Systems, Man, and Cybernetics, Part B: Cybernetics*. Her research interests include artificial intelligence, machine learning, graph-based relational data mining, smart environments, and robotics.

Aaron S. Crandall is a Ph.D. student in the School of Electrical Engineering and Computer Science at Washington State University, USA. He received his master's degree from Oregon Graduate Institute, USA, in 2005. His areas of interest include machine learning, evolutionary computation, and smart environments.

Brian Dale received his Bachelors of Environmental Design from the University of Colorado, USA in 2001 before going on to complete a Master of Architecture and Urbanism at the Architectural Association's Design Research Laboratory [AADRL] in 2007, where the work of his DRL team KNFRK was published in DRL TEN: A Design Research Compendium. Prior to moving to London, he co-directed Everything Possible, a collective for emerging artists, in which he showcased his own photographic and interactive video installations. Having spent time working for Richard Epstein Architects and AR7 Architects in Colorado, he currently designs for Zaha Hadid Architects in London, UK.

Christos Goumopoulos received his diploma and Ph.D. degrees in computer science from University of Patras, Hellas, Greece, in 1992 and 2000, respectively. Since 1992 he has been involved as member or leader of scientific and technology teams in several EU-funded R&D projects. Currently he is a cooperating professor in the Hellenic Open University and serves as a term-appointed assistant professor in the University of Patras and as teaching staff in the Technological Educational Institute of Patras, Greece. His research interests include software engineering, programming languages and compilers, resource scheduling, distributed computing, ubiquitous computing and awareness management, middleware and ontological knowledge representation.

Hani Hagras received the B.Sc. and M.Sc. degrees from the Electrical Engineering Department at Alexandria University, Egypt, and the Ph.D. degree in computer science from the University of Essex, UK. He is currently a professor in the Department of Computing and Electronic Systems, director of the Computational Intelligence Centre, and Head of the Fuzzy Systems research groups in the University of Essex, UK. His major research interests are in computational intelligence, notably fuzzy logic, neural networks, genetic algorithms, and evolutionary computation. His research interests also include ambient intelligence, pervasive computing, and intelligent buildings. He is also interested in embedded agents, robotics, and intelligent machines. Professor Hagras is a fellow IET and senior member of IEEE, chair of the IEEE CIS Senior Members Nomination Subcommittee, chair of the IEEE CIS Task Force on Intelligent Agents and chair of the IEEE CIS Task Force on Extensions to Type-1 Fuzzy Sets. He is a member of the IEEE Technical Committee of the Building Automation, Control and Management and the IEEE Fuzzy Systems Technical Committee. In addition, he is also a member of the executive committee of the IEEE Robotics and Mechatronics Professional Network.

Tobias Heinroth received his diploma degree in computer science from the University of Ulm, Germany, in 2007. He is currently pursuing his Ph.D. degree as a member of the Dialogue Systems Group at the Institute of Information Technology at the University of Ulm. His current research focus lies on exploring ways of managing spoken dialogues in intelligent environments. His general research interests include man–machine communication, ambient intelligent environments, navigation systems, and multimodal human–computer interaction.

Abdelsalam (Sumi) Helal is a professor at the Computer and Information Science and Engineering Department (CISE) at the University of Florida, USA. His research interests span the areas of pervasive computing, mobile computing, and networking and Internet computing. He directs the Mobile and Pervasive Computing Laboratory and the Gator Tech Smart House, an experimental home for applied pervasive computing research in the domain of elder care. Additionally, he is founder, president, and CEO of Phoneomena, Inc., a mobile application and middleware company, and founder and president of Pervasa, Inc., a University of Florida start-up focused on platform and middleware products for sensor networks.

Christoph Hölscher is an assistant professor at the University of Freiburg, Centre for Cognitive Science, Germany. He received his diploma in psychology from the Ruhr-Universität Bochum, Germany, in 1997, and his Dr. Phil. in psychology from the University of Freiburg, Germany, in 2000. He was a senior researcher and project manager in the IT industry from 2000 to 2003, when he re-joined academia in Freiburg, Germany. In addition to his post at the University of Freiburg he currently serves as an Honorary Senior Research Fellow at University College London, Bartlett School of Graduate Studies, UK.

Sajid Hussain is an assistant professor in the Jodrey School of Computer Science, Acadia University, Canada. He received a Ph.D. in electrical engineering from the University of Manitoba, Canada. Dr. Hussain is investigating intelligent and energy-efficient data dissemination techniques in sensor networks for ubiquitous and pervasive applications.

Obidul Islam is a software developer at IBM Ottawa Lab. He obtained his M.Sc. in computer science from Acadia University, Canada, in 2008. He was a research assistant at Jodrey School of Computer Science, Acadia University from 2006 to 2007.

Christophe Jacquet is an assistant professor at Supélec (École Supérieure d'Électricité), France. He holds an engineering degree and a Master of Science. He received a Ph.D. in computer science from the University of Paris-Sud, France, in 2006. His research interests include heterogeneity management and embedded systems, with applications to ambient intelligence.

Vikramaditya R. Jakkula received his master's degree from the School of Electrical Engineering and Computer Science at Washington State University, USA, in 2007. His areas of interest include machine learning, intelligent systems, data mining, and artificial intelligence.

Brian David Johnson is a consumer experience architect within Intel's Digital Home Group. His responsibilities include researching, defining, and mapping the public's experience with future products and services. Before joining Intel, he served as executive producer on several interactive television deployments in Scandinavia, Europe, and the United States for British Airways, The Discovery Channel, and New Line Cinema's *The Lord of the Rings*. Johnson holds a B.A. from the New School for Social Research, New York City, USA. He is the director of the feature films *POP and Haunters* and the author of the science fiction novels *Fake Plastic Love* and the forthcoming *This is Planet Earth*.

Gerard Joson is a Manila-born designer. He received his B.S. in Architecture from the University of the Philippines in 2003, and trained right after under JY+A, one of the few firms creating avant-garde design solutions within the

Philippine context. He continued his education with product, furniture and graphic design courses in 2004 at the Pratt Institute and School of Visual Arts in New York, USA, before earning his Masters in Architecture [DRL] in 2007 at the Architectural Association School of Architecture in London, UK. Back in the Philippines, he currently heads his family-owned development company and actively does consultantcy work for various design projects with his company joson_design. In mid-2009, he will be starting as a professor at the College of Architecture, University of the Philippines.

Youssef Kaddourah is a Ph.D. candidate at the University of Florida, USA. He joined the Harris Mobile and Pervasive Computing Laboratory in 2001 and contributed significantly to the developments of the Gator Tech Smart House. His research areas include indoor location tracking and positioning and geomatics.

Achilles D. Kameas received his Engineering Diploma (in 1989) and his Ph.D. (in 1995, in human–computer interaction), both from the Department of Computer Engineering and Informatics, University of Patras, Greece. He has also received formal education on Adult Education and on Open and Distance Education. Since 2003, he is an assistant professor with the Hellenic Open University, where he teaches software design and engineering. He is also R&D manager with Research Academic Computer Technology Institute (CTI), Greece, where he is the head of Research Unit 3 (Applied Information Systems) and the founder of DAISy (Designing Ambient Intelligent Systems) group. Since 2007 he is deputy dean of the School of Sciences and Technology (SST) of the Hellenic Open University and Director of the e-Comet Lab (Educational Content, Methodologies and Technologies Lab). He has participated as researcher, engineer, group leader, or scientific coordinator in several EU and national R&D projects, such as e-Gadgets, Plants, Social, Astra, and Atraco. His current research interests include architectures, languages, ontologies and tools for ubiquitous computing systems, engineering of ubiquitous computing applications, and distributed component-based systems. He is a voting member of IEEE, IEEE CS, ACM, and ACM SIGCHI. He is a member of Technical Chamber of Greece, Hellenic AI Society and Hellenic Society for the application of ICT in Education.

Jeffrey King obtained his Ph.D. from the University of Florida, USA, in 2007. His main area of research is sensor platform operating systems. He was a research assistant in the Harris Mobile and Pervasive Computing Laboratory from 2004 to 2007. He is one of the key contributors to the Gator Tech Smart House.

Wolfgang Minker is a professor at the University of Ulm, Institute of Information Technology, Germany. He received his Ph.D. in engineering science from the University of Karlsruhe, Germany, in 1997 and his Ph.D. in computer science from the University of Paris-Sud, France, in 1998. He was a researcher at the Laboratoire d'Informatique pour la Mécanique et les Sciences de l'Ingénieur (LIMSI-CNRS), France, from 1993 to 1999 and a member of the scientific staff at DaimlerChrysler, Research and Technology, Germany, from 2000 to 2002.

Ioannis Orfanos is a PhD candidate in the Department of Architectural Technology at the National Technical University of Athens. In 2004 received his Diploma in Architecture, followed by the theoretical post-graduate course Design-Space-Culture in the School of Architecture of NTUA. He holds a MArch in Architecture and Urbanism from Design Research Laboratory in Architectural Association School of Architecture, London, UK. He is a registered architect in ARB(UK) and TCG(Gr). He has worked as self-employed architect, for Kohn Pedersen Fox Associates in London and MYAA Architects in Barcelona.

Günther Palm studied mathematics at the Universities of Hamburg and Tübingen, Germany. After his graduation he worked at the Max-Planck-Institute for Biological Cybernetics in Tübingen on the topics of non-linear systems, associative memory, and brain theory. In 1983/1984, he was a fellow at the Wissenschaftskolleg in Berlin, Germany. From 1988 to 1991 he was professor for Theoretical Brain Research at the University of Düsseldorf, Germany. Since then he is professor for Computer Science and Director of the Institute of Neural Information Processing at the University of Ulm, Germany. His research topics in computer science include information theory and applications of artificial neural networks in speech, vision, robotics, sensor-fusion, and pattern recognition.

Olga Pantelidou received her Diploma in Architectural Engineering from Aristoteleio University of Thessaloniki, Greece, in 1998. Currently, she is a Ph.D. candidate at the School of Architecture at the National Technical University of Athens, Greece, and a graduate student at the Yale School of Architecture's MED program, New Haven, CT, USA. She holds two master degrees, one in Information Systems (University of Macedonia of Economics & Social Sciences, Thessaloniki, Greece, 2001), and another in Architecture-Space Design (National Technical University of Athens, 2004). She taught at the University of Thessaly, Greece, in the Department of Planning and Regional Development from 2004 to 2006. In 2004, she participated in the 9th International Exhibition of Architecture, "Bienalle di Venezia."

Stefan Scherer was born in Feldkirch, Austria, in 1983. He received his Diploma in Computer Science in 2006 from the University of Ulm, Germany. Since 2006, he studies as a Ph.D. student in the Institute of Neural Information Processing at the University of Ulm. His research interests, which are being developed in his doctoral thesis include affective computing, multiple classifier systems, pattern recognition, and feature selection.

Friedhelm Schwenker studied mathematics and computer science at the Universities of Osanbrück and Hagen, Germany. After his graduation he worked at the Faculty of Mathematics and Computer Science, University of Osnabrück, Germany, and at the Vogt-Institute for Brain Research, University of Düsseldorf, Germany. Since 1992 he is a senior researcher/lecturer at the Institute of Neural Information Processing, University of Ulm, Germany. His research topics are artificial neural networks, machine learning, pattern recognition, signal processing, multiple classifier systems, sensor fusion, approximation theory, and applied statistics.

Michael Weber holds a Ph.D. in computer science from the University of Kaiserslautern, Germany. After a number of years in industry, working on parallel and multimedia systems, he joined the University of Ulm, Germany, as a professor for computer science in 1994 and was appointed director of the Institute of Media Informatics in 2000. He has authored and co-authored more than 100 peer-reviewed contributions, edited three books and written a textbook. He has led projects funded by the state of Baden-Württemberg, by the German Ministry for Education and Research (BMBF), by the European Commission and by industrial partners. His current research interests include mobile and ubiquitous computing systems and human–computer interaction.

Gregor Wilbertz studies psychology at the University of Freiburg, Germany. Since 2005 he has been a student assistant in the research project "Architectural Design and Wayfinding Cognition" at the Centre for Cognitive Science, University of Freiburg, Germany. He has been involved with all research phases of several wayfinding studies, both in real-world settings and for computer-based simulation studies.

Katharine S. Willis is an EU Marie Curie Research Fellow on the MEDIACITY project, Bauhaus-Universität Weimar, Germany. Prior to this she was doctoral researcher on the spatial cognition program at the University of Bremen, Germany, funded by a DAAD scholarship. She received a Master in Architecture (commendation) from the Bartlett School of Architecture, UCL, London, England, in 2000 and her Diploma in Architecture in 1998.

Pavlos Xanthopoulos is a PhD candidate at the Architectural Technology Department of the National Technical University of Athens, Greece. In 2004 he received his Diploma in Architecture from the the School of Architecture of the NTUA. He holds a MArch in Architecture and Urbanism from Design Research Laboratory in Architectural Association School of Architecture, London, UK. He is a registered architect in ARB(UK) and TCG(Gr). In 2004 he cofounded otn/studio, a young professional architectural design practice. Pavlos has worked as an architect for Zaha Hadid Architects. His work has been published in the DRL TEN: A Design Research Compendium and the “emerging technologie” and self-sufficient housing by Actar.

Jinchun Xia is a visiting professor in the Department of Computer Engineering at Sanjose State University, USA. She received her Ph.D. in Computer Science from Iowa State University, USA. She also obtained her M.S. in cryptography from Southwest Jiaotong University, China, in 2001. Her research focuses on performance engineering, service-oriented computing, distributed software engineering and net-centric computing. She is a member of IEEE.

Hen-I Yang is a post-doctoral fellow in the Department of Computer Science at Iowa State University, USA. He obtained his Ph.D. in Computer Engineering from the Mobile and Pervasive Computing Laboratory at University of Florida, USA. His research interests include system safety and reliability, programming models, and system and middleware support for pervasive and mobile computing. He is a member of IEEE and ACM.

Hicham EL-Zabadani obtained his Ph.D. from the University of Florida, USA, in 2006. His main area of research is self-sensing spaces. El-Zabadani was a main contributor to the Gator Tech Smart House. He was a research assistant in the Harris Mobile and Pervasive Computing Lab from 2002 to 2006.

Introduction

Wolfgang Minker
Institute of Information Technology, Ulm University, Ulm, Germany
wolfgang.minker@uni-ulm.de

Michael Weber
Institute of Media Informatics, Ulm University, Ulm, Germany
michael.weber@uni-ulm.de

This book highlights recent trends and important issues contributing to the realization of the ambient intelligence vision, where physical space becomes augmented with computation, communication, and digital content, thus transcending the limits of direct human perception. The focus is placed on advanced inhabitable intelligent environments including mechanisms, architectures, design issues, applications, evaluation, and tools.

The book is based on a selected subset of papers from the IET International Conference on intelligent environments (IE 07) held in Ulm, Germany. This conference has been the third in the highly successful intelligent environments (IE) conference series where the first conference (IE 05) took place in Colchester, UK, in June 2005 and the second conference took place in Athens, Greece, in July 2006. In April 2007, the conference series was awarded the Knowledge Network Award by the Institution of Engineering and Technology (IET) as the conference series was perceived to be emerging as the strongest international multi-disciplinary conference in the field. The conference brings together the contributions of different intelligent environments disciplines to form a unique international forum that will help to create new research directions in the intelligent environments area while breaking down barriers between the different disciplines. In addition, the conference provides a leading edge forum for researchers from industry and academia from across the world to present their latest research and to discuss future directions in the area of intelligent environments.

The IE 07 conference programme featured 91 papers from more than 23 different countries representing the 6 continents. Of these nine were invited for publication in this book along with a paper by an invited speaker, i.e. a total of 10 papers. All conference papers were extended and revised before they were submitted as book chapters. Each chapter has subsequently been reviewed by at least two reviewers and further improved on the basis of their comments.

We would like to thank all those who contributed to and helped us in preparing the book. In particular we would like to express our gratitude to the following reviewers for their valuable comments and criticism on the submitted drafts of the book chapters: Elisabeth André, Hakan Duman, Hans Dybkjær, Kjell Elenius, Michael Gardner, Franz Hauck, Sumi Helal, Anne Holohan, Sajid Hussain, Rosa Iglesias, Nicos Komninos, Antonio Lopez, Michael McTear, Anton Nijholt, Angelica Reyes, Albrecht Schmidt, Abdulmotaleb El Saddik, and Roger Whitaker. We are also grateful to Kseniya Zablostkaya and Sergey Zablotskiy at the Institute of Information Technology at the University of Ulm for her support in editing the book.

In the following we give an overview of the book contents by providing excerpts of the chapter abstracts. Very roughly we may divide the chapters into the following categories although many chapters address aspects from more than one category and all chapters deal with intelligent environment aspects.

- Pervasive computing (Chapters 1–3);
- Human–computer interaction (Chapters 4–6);
- Context awareness (Chapters 7–8);
- Architecture (Chapters 9–11).

Pervasive computing: Chapters 1–3 deal with issues in the area of pervasive computing.

In Chapter 1 Helal et al. present an assistive environment for health-care and well-being services to elderly people (Helal et al., 2009). The demand for senior-oriented devices and services will significantly increase in the near future. Assistive environments provide support and compensate for age-related impairments. Pervasive computing environments, such as smart homes, bundle assistive technologies and specially designed architectural and home furnishing elements. However, to be commercially viable, a system should allow the technology to be easily utilized and be introduced in a plug-and-play fashion. As an example for assistive environments, the authors present a residential home for elderly people.

Johnson explores in Chapter 2 consumer experience architecture as a practice and a methodology for developing products and services so that they fit intuitively into the lives of consumers (Johnson, 2009). He draws on recent experiences at Intel, where this framework has directly been applied to the development of personal technology devices. The chapter dismantles the consumer experience architecture into its essential components, exploring real-world examples and illustrations. The reader is challenged to expand current development practices by looking towards science fiction or other cultural inputs as possible laboratories or inspirations for future designs.

According to Goumopoulos et al. (Chapter 3) artifacts will have a dual self in the forthcoming Ambient Intelligence environments: artifacts are objects with physical properties and they have a digital counterpart accessible through a network (Goumopoulos et al., 2009). An important characteristic may be the merging of physical and digital space (i.e. tangible objects and physical environments are acquiring a digital representation), still, people's interaction with their environment will not cease to be goal-oriented and task-centric. However, ubiquitous computing technology will allow people to carry out new tasks, as well as old tasks in new and better ways.

Human–computer interaction: Chapters 4–6 deal with issues in the area of human–computer interaction in intelligent environments.

In Chapter 4 Jacquet et al. propose a ubiquitous information system providing personalized information to mobile users, such as in airports and train stations (Jacquet et al., 2009). The goal is to perform a selection among the set of available information items, so as to present, in a multimodal way, only those relevant to people located at proximity. A device will provide information to a user only if one of its output modalities is compatible with one of the user's input modalities. The proposed agent architecture is based on an alternative to traditional software architecture models for human–computer interaction.

The trend in affective computing currently aims towards providing simpler and more natural interfaces for human–computer interaction. The computer should be able to adapt its interaction policies to the user's emotional status. Scherer et al. investigate in Chapter 5 the performance of an automatic emotion recognizer using biologically motivated features (Scherer et al., 2009). Single classifiers using only one type of features and multi-classifier systems utilizing all three types are examined using two classifier fusion techniques. The performance is compared with earlier work as well as with human recognition performance. Using simple fusion techniques could improve the performance significantly.

In Chapter 6 Willis et al. investigate the nature of spatial knowledge acquisition in an environmental setting (Willis et al., 2009). The authors use a task where the participants have learnt the environment using spatial assistance, either from a map or from a mobile map. Results of an empirical experiment which evaluated participants spatial knowledge acquisition for orientation and distance estimation tasks in a large-scale urban environmental setting are outlined. The experiments showed that mobile map participants performed worse in distance estimation tasks than map participants, especially for complex routes.

Context awareness: Chapters 7–8 deal with context awareness in intelligent environments.

In Chapter 7 Hussain and Islam present a genetic algorithm to generate balanced and energy-efficient data aggregation spanning trees for wireless sensor networks (Hussain and Islam, 2009). These networks are commonly used in various ubiquitous and pervasive applications. Due to limited power resources, the energy-efficient communication protocols and intelligent data dissemination techniques are needed. Otherwise, the energy resources will deplete drastically and the network monitoring will be severely limited. In a data aggregation environment, the gathered data are highly correlated and each node is capable of aggregating any incoming messages to a single message and reduce data redundancy.

The objective of the research described by Jakkula et al. in Chapter 8 is to identify temporal relations among daily activities in a smart home to enhance prediction and decision-making with these discovered relations, and to detect anomalies (Jakkula et al., 2009). The authors hypothesize that machine learning algorithms can be designed to automatically learn models of resident behavior in a smart home. When these are incorporated with temporal information, the results can be used to detect anomalies. This hypothesis is validated using empirical studies based on the data collected from real resident and virtual resident data.

Architecture: Chapters 9–11 address architectural issues, both in terms of computer architecture (middleware) and bulidings and structures.

Service-oriented architecture, addressed by Yang et al. in Chapter 9, has established itself as a prevailing software engineering practice in recent years and extends to the domain of pervasive computing (Yang et al., 2009). The proposed solution for building fault-resilient pervasive computing systems consists of two parts: First, the virtual sensor framework improves the availability of basic component services. Second, an architecture for performing service composition can efficiently model, monitor and re-plan this process. To create a comprehensive solution, these two parts have to work hand in hand during the entire life cycle of pervasive services.

Chapter 10 by Dale et al. describes a system of parametric-networked urbanism that explores the integration of adaptive spaces according to cultural, social and economic dynamics (Dale et al., 2009). The goal of the research was to explore new forms of urbanism corresponding to criteria of parametric design and further the development of a proposal about the London area. Embedded with self-learning behavioral and responsive systems, the project allows for an intelligent choreography of soft programmatic spaces to create new leisure experiences, negotiating the changing effects of time, weather, programmatic, and crowd dynamical inputs, extending parametric processes to drive urban performance.

Pantelidou introduces in Chapter 11 the concept of the totality of space, defining it as a corporation's bounded spaces and the connections between them (Pantelidou, 2009). This concept expresses itself in the evolution of banking in the twentieth century. The chapter argues the importance of revealing and understanding the characteristics of the totality of space, which are inherent to the banking industry's spatial thought, thus allowing architects to bring the knowledge of their field and participate in a design/planning process of directing its possible future forms.

We believe that jointly this collection of chapters provides a good picture of how far we are today within the AmI vision and of the important challenges ahead. On this background we hope that computer scientists, engineers, architects and others who work in the broad area of intelligent environments, no matter if from an academic or industrial perspective, may benefit from the book and find it useful to their own work. Graduate students and Ph.D. students focusing on AmI-related topics may also find the book interesting and profit from reading it.

References

Dale, B., Orfanos, I., Xanthopoulos, P., and Joson, G. (2009). Intravein – Parametric Urbansim. In *"Advanced Intelligent Environments"*. Springer, This Edition.

Goumopoulos, C., Kameas, A., Hagras, H., Callaghan, V., Heinroth, T., and Weber, M. (2009). An Architecture that Supports Task Centered Adaptation in Intelligent Environments. In *"Advanced Intelligent Environments"*. Springer, This Edition.

Helal, A. S., King, J., Bose, R., EL-Zabadani, H., and Kaddourah, Y. (2009). Assistive Environments for Successful Aging. In *"Advanced Intelligent Environments"*. Springer, This Edition.

Hussain, S. and Islam, O. (2009). Genetic Algorithm for Energy Efficient Trees in Wireless Sensor Networks. In *"Advanced Intelligent Environments"*. Springer, This Edition.

Jacquet, C., Bourda, Y., and Bellik, Y. (2009). Multimodal Presentation of Information in a Mobile Context. In *"Advanced Intelligent Environments"*. Springer, This Edition.

Jakkula, V. R., Crandall, A. S., and Cook, D. J. (2009). Enhancing Anomaly Detection Using Temporal Pattern Discovery. In *"Advanced Intelligent Environments"*. Springer, This Edition.

Johnson, B. (2009). Do Digital Homes Dream of Electric Families? Consumer Experience Architecture as a Framework for Design. In *"Advanced Intelligent Environments"*. Springer, This Edition.

Pantelidou, O. (2009). The Totality of Space. In *"Advanced Intelligent Environments"*. Springer, This Edition.

Scherer, S., Schwenker, F., and Palm, G. (2009). Classifier Fusion for Emotion Recognition from Speech. In *"Advanced Intelligent Environments"*. Springer, This Edition.

Willis, K. S., Hölscher, C., and Wilbertz, G. (2009). Understanding Mobile Spatial Interaction in Urban Environments. In *"Advanced Intelligent Environments"*. Springer, This Edition.

Yang, H.-I., Bose, R., Helal, A. S., Xia, J., and Chang, C. K. (2009). Fault-Resilient Pervasive Service Composition. In *"Advanced Intelligent Environments"*. Springer, This Edition.

Chapter 1

ASSISTIVE ENVIRONMENTS FOR SUCCESSFUL AGING

Abdelsalam (Sumi) Helal, Jeffrey King, Raja Bose, Hicham EL-Zabadani, Youssef Kaddourah

Department of Computer and Information Science and Engineering
University of Florida, Gainesville, Florida, USA

{helal, jck, rbose, hme}@cise.ufl.edu, kaddoura@ufl.edu

Abstract With nearly 80 million baby boomers in the United States just reaching their sixties, the demand for senior-oriented devices and services will explode in the coming years. Managing the increasing health-care costs for such a population requires developing technologies that will allow seniors to maintain active, independent lifestyles. Pervasive computing environments, such as smart homes, bundle assistive technologies and specially designed architectural and home furnishing elements provide health-care and well-being services to its residents. However, for such environments to be commercially viable, we require a system that allows technology to be easily utilized and included as it enters the market place. Also we require new technology to be introduced in a plug-and-play fashion, and applications that are developed by programmers, not system integrators. The Gator Tech Smart House, a full-size, free-standing residential home located in the Oak Hammock Retirement Community in Gainesville, Florida, is an example of this kind of assistive environment. It uses the Atlas sensor network platform, an enabling technology that combines a hardware platform and software middleware, making the Gator Tech Smart House a truly programmable pervasive computing space.

Keywords: Assistive technology; Sensor networks; Ubiquitous service composition; Pervasive computing; Sensor platform.

1. Introduction

Research groups in both academia and industry have developed prototype systems to demonstrate the benefits of pervasive computing in various

W. Minker et al. (eds.), *Advanced Intelligent Environments*,
DOI 10.1007/978-0-387-76485-6_1,

application domains. These projects have typically focused on basic system integration-interconnecting sensors, actuators, computers, and other devices in the environment. Unfortunately, many first-generation pervasive computing systems lack the ability to evolve as new technologies emerge or as an application domain matures. Integrating numerous heterogeneous elements is mostly a manual, ad hoc process. Inserting a new element requires researching its characteristics and operation, determining how to configure and integrate it, and tedious and repeated testing to avoid causing conflicts or indeterminate behavior in the overall system. The environments are also closed, limiting development or extension to the original implementers.

To address this limitation, the University of Florida's Mobile and Pervasive Computing Laboratory is developing programmable pervasive spaces in which a smart space exists as both a runtime environment and a software library (Helal, 2005). Service discovery and gateway protocols automatically integrate system components using generic middleware that maintains a service definition for each sensor, actuator, and device in the space. Programmers assemble services into composite applications, which third parties can easily implement or extend.

The use of service-oriented programmable spaces is broadening the traditional programmer model. Our approach enables domain experts – for example, health professionals such as psychiatrists or gastroenterologists – to develop and deploy powerful new applications for users.

In collaboration with the university's College of Public Health and Health Professions, and with federal funding as well as donations and gifts, we have created a programmable space specifically designed for the elderly and dis-

Figure 1. Front view of the Gator Tech Smart House.

abled. The Gator Tech Smart House (GTSH), shown in Figure 1, located in the Oak Hammock Retirement Community in Gainesville, Florida, is the culmination of more than 6 years of research in pervasive and mobile computing. The project's goal is to create assistive environments that can provide special services to the residents to compensate for cognitive, mobility, and other age-related impairments (Helal et al., 2005).

2. Assistive Services in the Gator Tech Smart House

Figure 2 shows most of the "hot spots" that are currently active or under development in the Gator Tech Smart House. An interactive 3D model (GTSH, 2007) provides a virtual tour of the house with up-to-date descriptions of the technologies arranged by name and location. This section will describe several of the major services and features provided in this assistive environment.

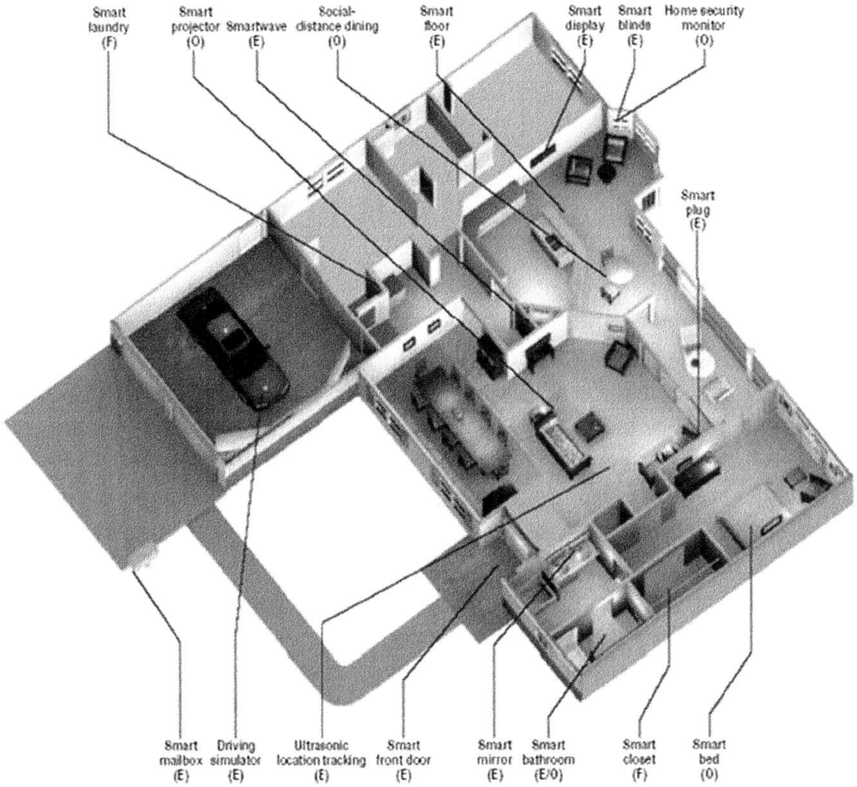

Figure 2. Gator Tech Smart House floorplan. The project features numerous existing (E), ongoing (O), and future (F) "hot spots" located throughout the premises.

2.1 Entry Assistant

The front door area of the Gator Tech Smart Houses makes use of several devices and services that together comprise the entry assistant. A radio-frequency identification (RFID) system built into the wall of the entranceway recognizes residents as they approach the house by means of passive RFID tags attached to their key rings. Two devices, an electronic deadbolt and an automatic door opener (Figure 3), work together to allow the residents access to the house and to secure the premises when the door is closed.

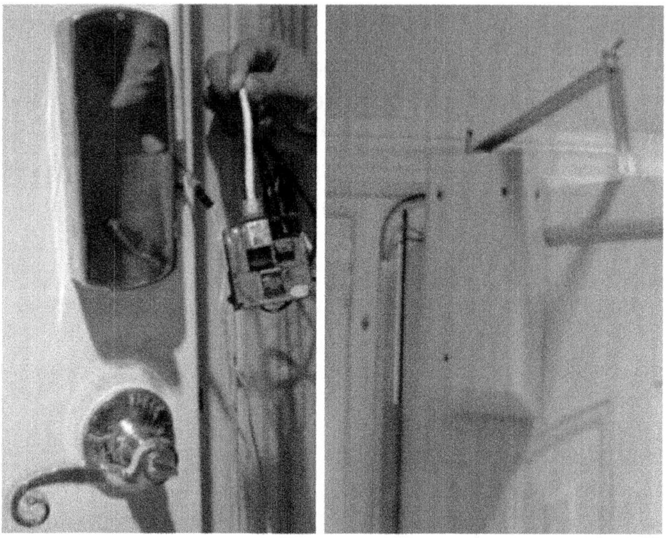

Figure 3. Entry assistant front door, with electronic deadbolt (*left*) and door opener (*right*).

The doorbell of the house connects to the smart space. This allows the Gator Tech Smart House to easily adapt the notification system to the needs of its residents. For example, a visual indicator such as a flashing light can be provided to a resident with a hearing impairment. The doorbell also triggers the door view service – a small video camera built into the peephole of the door. The video is automatically transmitted to the monitor nearest the resident in the house. Access to the house can be granted with a voice command or the resident may provide the visitor with audio or text messages using the speakers or LCD display built into the entranceway.

While the entry assistant provides several smart services for the resident, an important note for this and our other applications is that the "dumb" functionality of devices is never removed. The automatic door opener we chose is free swinging, meaning the door can be opened or closed by hand. The electronic deadbolt, while containing an internal motor for automatic control, also has a normal key interface outside and knob inside. Residents are not forced to do things the "new" way.

2.2 Location Tracking and Activity Monitoring

Location tracking is a fundamental service in a pervasive computing environment such as a smart house. The location of residents in the house can trigger or halt certain applications, affect various notification systems in the environment, and can be used to ascertain data about the health of the residents in terms of daily activity or detecting falls.

The Gator Tech Smart House supports a variety of location tracking technologies. The original technology used, carried over from our in-lab prototype house, is an ultrasonic tag-based system. Each resident is given a pair of transceivers to wear, and transmissions between these devices and transceivers in the ceiling are able to triangulate each resident.

While there are several benefits to such a system, such as the ease of multi-resident tracking, and the ability to detect the direction each resident is facing, the major drawback to this system is that it requires active participation by the residents: for the house to locate them, they must remember to put on the transceivers, ensure that the batteries are charged, etc.

The primary location tracking system used in the Gator Tech Smart House is the smart floor (Kaddourah et al., 2005). The flooring for the entire house consists of residential-grade raised platform. Each platform is approximately one square foot, and underneath each we installed a force sensor (Figure 4). Unlike the ultrasonic tag method, the smart floor requires no attention from the residents, and unlike some other unencumbered tracking systems, there are no cameras that invade the residents' privacy. This allows for constant but inoffensive monitoring throughout the day and night, even in areas such as the bathroom. Figure 5 shows an example of this tracking.

Applications that make use of the smart floor service include the house's notification system. Alerts can be sent to the video or audio device nearest the resident. The entertainment system makes use of location information by following the resident throughout the house, turning off the television in one room and turning it on in another. More importantly, the activity monitor is able to record a resident's typical amount of movement in a day. If a significant decrease in activity is detected, the house is able to automatically notify caregivers.

Currently we are working to further improve our location tracking system. We are investigating the use of vibration sensors located at certain points in the house to replace the full coverage of force sensors. While this would be more expensive in terms of device cost, the time necessary to deploy the solution is significantly less, allowing for a packaged solution. Additionally, it would

Figure 4. Tile of the smart floor.

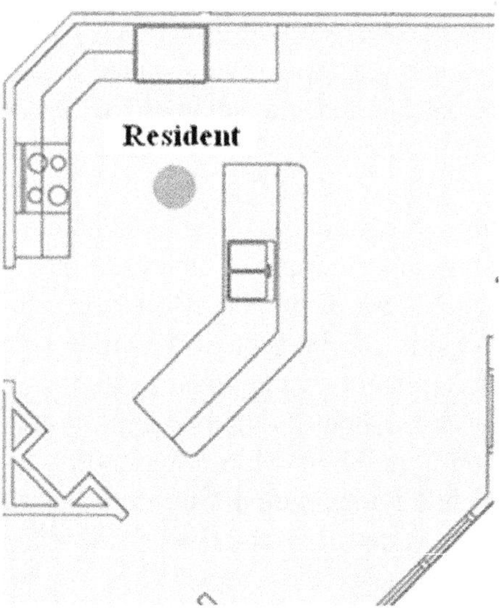

Figure 5. Visual location tracking application.

allow smart floor technology to be installed in any home, not just those with raised flooring.

We are also looking at improving the activity monitoring support by including tracking technology in more than the floor. For example, similar force sensors in the bed can be used to detect when residents are sleeping. Variations in sleep patterns would be of interest to the residents and caregivers.

2.3 SmartWave

The SmartWave (Figure 6) is a collection of devices and services that facilitates meal preparation in the Gator Tech Smart House (Russo et al., 2004). A standard microwave oven was modified to allow computer control of the cooking process. An RFID reader in the cabinet below the microwave allows appropriately tagged frozen meals to be placed in front of the device and recognized by the smart house.

Figure 6. The SmartWave meal preparation assistance system.

The resident will be provided with any necessary instructions to ready the meal for cooking (remove film, stir ingredients, etc.). The SmartWave will set power levels and cooking times automatically. This technology assists a variety of residents, such as those with visual impairments who are unable to read the fine print on the frozen meals. Once the meal is ready, a notification will be sent to the resident, wherever he/she is in the house.

2.4 Real-World Modeling for Remote Monitoring and Intervention

An assistive environment such as the Gator Tech Smart House should provide tools and services that benefit both the residents and the residents' caregivers. In many cases, however, caregivers will be living off-site. Caregivers include the residents' adult sons and daughters, or contracted healthcare providers. In either case, situations will arise where the caregivers will need a remote presence in the house.

To support this remote presence, we developed a number of research projects under the heading self-sensing spaces. A smart house should be able to recognize the devices and services it has available, interpret their status, and generate a model of the space (Figure 7). It should also recognize the residents and their activities, and include a representation of these in the model.

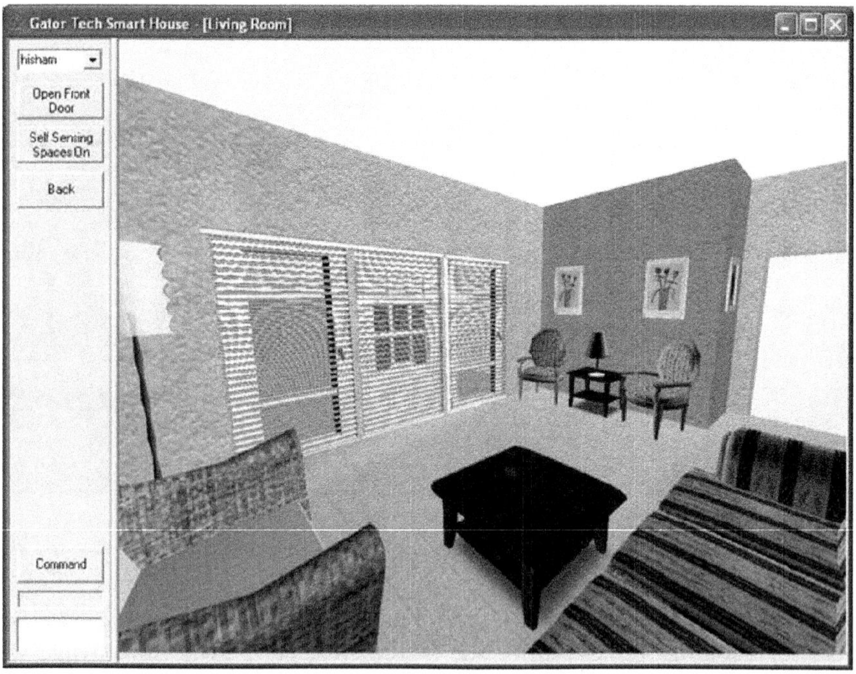

Figure 7. Real-world model of the smart house, provided to remote caregivers.

2.4.1 Smart Plugs. We first broached this issue of allowing the house to recognize installed devices with our smart plug project (El-Zabadani et al., 2005). Smart plugs include an RFID reader behind each electrical wall socket in the house (Figure 8, left). Each electrical device was then given an RFID tag that indicated what the device was and the commands it could be issued (Figure 8, right). This system allows the Gator Tech Smart House to detect devices as they enter or leave the space. A graphical model of the house is updated, providing remote caregivers with an accurate view of the capabilities of the house. In addition to just providing an image of the space, the system also allows remote caregivers to drive operation of devices. For example, if the caregiver notices that temperatures are climbing, they can click on fans to turn them on.

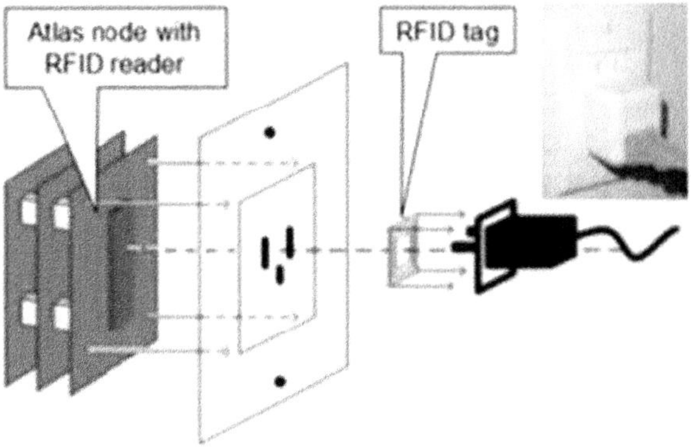

Figure 8. Smart plug deployed behind an electrical socket.

2.4.2 PerVision. While the smart plug system is able to detect active objects, we also require a system to detect passive objects such as furniture. The first iteration of this project, PerVision (El-Zabadani et al., 2006), made use of cameras and RFID to recognize and extract information about passive objects in the environment (Figure 9).

Before a passive object such as a chair or table was brought into the house, it was labeled with an RFID tag identifying certain characteristics about it: shape, volume, bounding box, color hues, etc. RFID readers deployed by the doors would detect items as they enter or leave the space. The PerVision system then used a series of cameras throughout the house to run image recognition techniques to determine the location of these objects as they were deployed and moved throughout the house. The computer vision techniques were assisted by information from the RFID tags and from the smart floor.

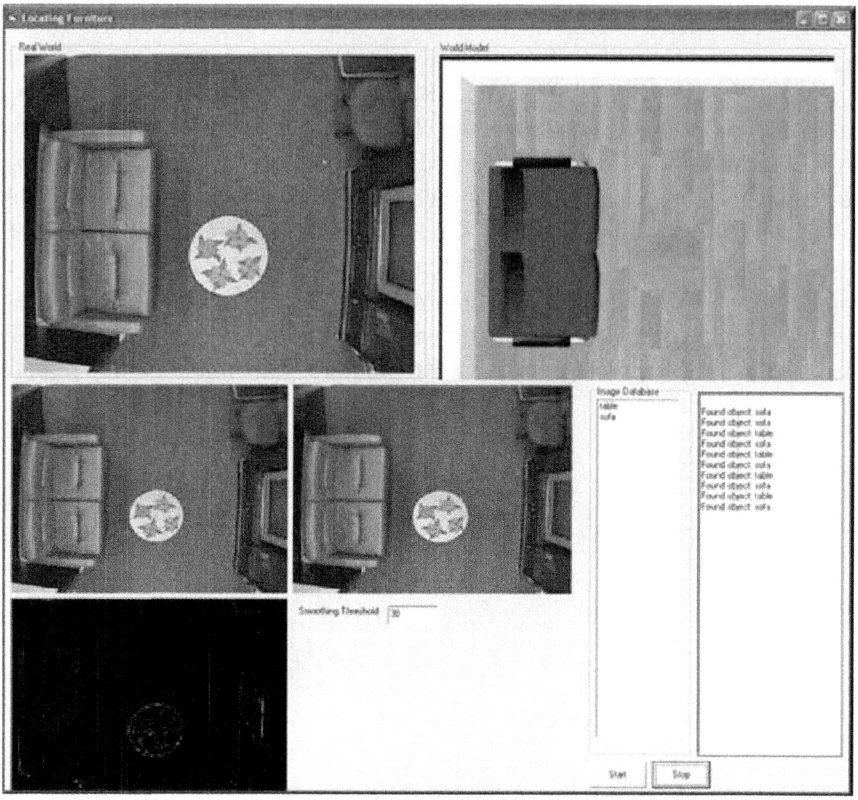

Figure 9. PerVision dynamic object recognition and extraction.

2.4.3 SensoBot. Although the PerVision system showed some success, we also required a less intrusive method of detecting objects. SensoBot (Figure 10) is a Roomba-based sensor platform that is able to physically map a space and detect the RFID-tagged objects in it. Not only did this provide a camera-less object detection method but it also generates a floor map of the space that feeds into other aspects of the self-sensing spaces project, such as the 3D model for remote caregivers.

In addition to its mapping and object recognition techniques, the SensoBot project proved useful in enabling many other services. Having a mobile sensor platform reduces the need for a space densely packed with sensors. This dramatically improves the cost-effectiveness of a smart space. Additionally, the mobility aspect can directly benefit residents: if a fall is detected, an emergency button-equipped SensoBot can be sent to the resident's location, instead of the system relying on the resident to always carry an emergency bracelet or pendant.

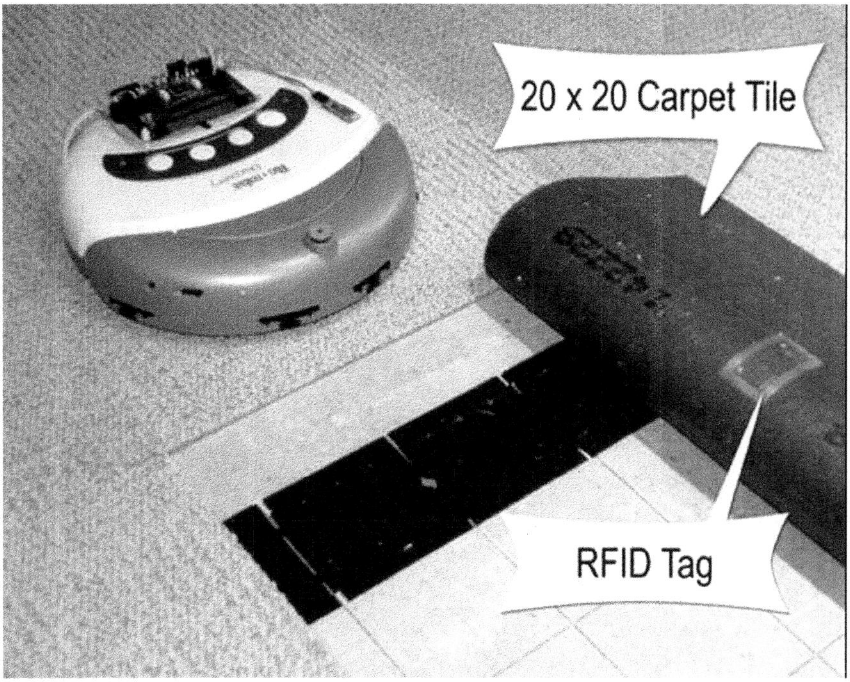

Figure 10. SensoBot recognizing an object in the carpet.

2.4.4 Smart Phone. The smart phone (Figure 11) is the "magic wand" of the Gator Tech Smart House (Helal and Mann, 2003). It is the master interface for all of the applications and devices provided by the assistive environment. It is a multimodal interface, offering traditional text menus, an icon-based interface, or voice recognition.

With the smart phone, the resident is able to control the comfort and convenience settings of the house, such as setting the desired temperature or adjusting the window blinds. The user can see the view from the front door's peephole camera, open the door, or lock the house at night. By connecting with the health-care applications in the house, the user can order prescription refills, or buy groceries online, with automatic feedback about any dietary restrictions.

The smart phone is also an integral part of the Gator Tech Smart House's medicine reminder service. Messages or other alerts are sent to the resident, reminding them to take their medication as needed. The phone includes a barcode scanner, which the resident uses to scan the bottles or cases of medicine. This allows the resident and the system to ensure that the correct medicine was taken at the correct time.

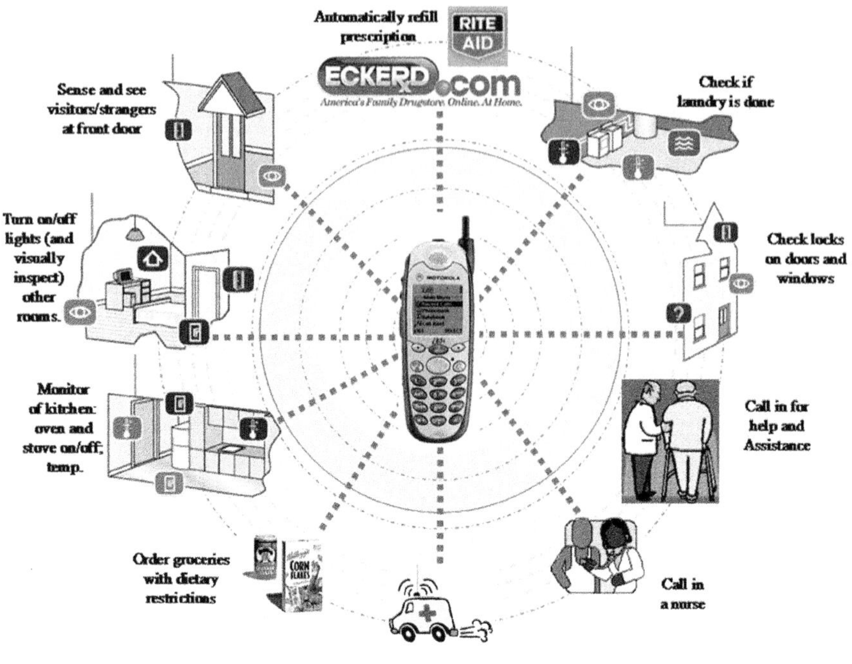

Figure 11. Smart phone as primary interface for the Gator Tech Smart House.

3. Technological Enablers for the Gator Tech Smart House

Pervasive computing systems such as the Gator Tech Smart House differ from traditional computing platforms. They are used in one's living and working space, and their services are heavily integrated with one's limitations, schedules, and preferences. These systems are obviously more intimate than traditional computing, and have a deep impact on peoples' daily lives and activities.

From our experiences during creation of our in-lab Matilda Smart House, as well as surveying various prototype systems and related publications, we noted certain critical properties that characterize pervasive computing systems:

- The system holds the capability to interact with the physical world. It is heavily dependent on hardware devices, including pure electric and mechanical devices without interfaces to computing systems.
- Services and applications are highly personalized and customized. Many of these customizations are achieved through the use of context and personal preferences. Many services require prediction of human intention, while others require personalized, non-traditional modalities.

- Failure is the norm during operation. The sheer number of devices and entities involved, the dynamicity and openness of the system, the diversity and heterogeneity of the devices are all contributing factors.
- The system is highly distributed. Instead of a well-defined and integrated configuration, it is most likely to be a federation of entities collaborating to achieve the common goal. This also ensures that the system needs to address scalability, complexity of the system organization, and administrative issues. It also requires the system architecture, and the interface to various entities, be standardized and open.

Addressing these issues in a project the size of the Gator Tech Smart House required creating a new platform on which large pervasive computing environments could be built. We first require a platform to connect numerous and heterogeneous sensors, actuators and other devices to the services and applications that will monitor and control the space. But connecting sensors and actuators to applications implies more than simply physically coupling these devices to a computer platform. Connecting devices with applications means providing some mechanism for the applications to make use of devices and services directly, instead of accessing some I/O resource on a machine that happens to be wired to a particular device. Beyond dealing with resource allocations, connecting applications and devices means eliminating the need for those applications to know the low-level information (voltages, control codes, etc.) to drive the devices.

To create the Gator Tech Smart House, we developed a generic reference architecture applicable to any pervasive computing space. As Figure 12 shows, the architecture contains separate physical, sensor platform, service, knowledge, context management, and application layers. This architecture is implemented as the Atlas middleware, and the Atlas sensor network platform.

3.1 Middleware

The Atlas middleware is a service-oriented platform that can "convert" the various sensors and actuators in a pervasive computing environment into software services. It is responsible for obtaining this service representation from connected devices, and for managing the services in such a way that applications are easily able to obtain and use the services and associated knowledge.

Two layers of the Atlas architecture, the services layer and the application layer, comprise the majority of the middleware. In our implementation, the services layer is built on top of the OSGi framework (Maples and Kriends, 2001). It holds the registry of the software service representation of all sensors

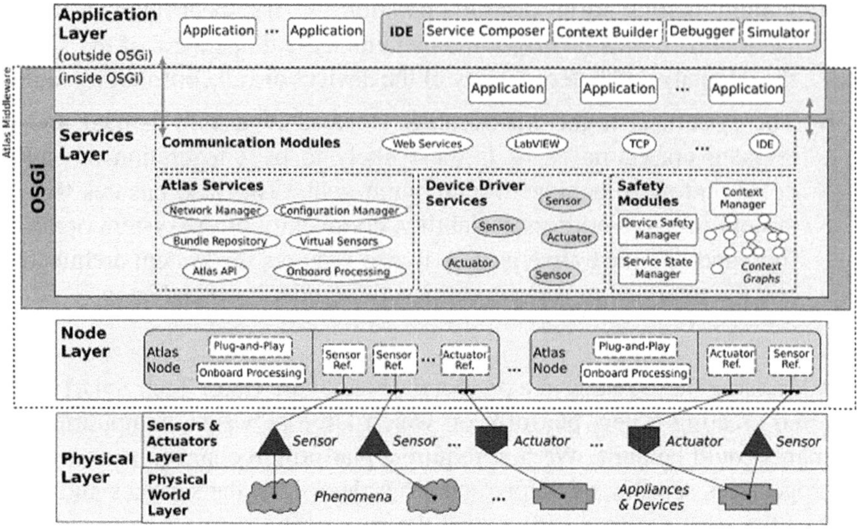

Figure 12. The Atlas reference architecture for programmable pervasive spaces.

and actuators connected to the hardware nodes. The layer provides the service discovery, composition, and invocation mechanisms for applications to locate and make use of particular sensors or actuators. It also supports enhancement of reliability and automatic service composition of sensor services.

OSGi (Figure 13) is a Java-based framework that provides a runtime environment for dynamic, transient service modules known as bundles. It provides functionalities such as life cycle management as well as service registration and discovery that are crucial for scalable composition and maintenance of applications using bundles. Designed to be the "universal middleware," OSGi enables service-oriented architectures, where decoupled components are able to dynamically discover each other and collaborate. OSGi is synergistic to pervasive computing (Lee et al., 2003), and is a key component of the Atlas middleware, hosting the majority of the software modules.

OSGi bundles are small programs consisting of three main source components: the interface, the implementation, and the OSGi activator. The interface represents a service contract, which describes the external behavior of and available services provided by the bundle. A bundle can provide different services by offering multiple interfaces. The implementation realizes the behavior defined by the interface. The activator implements an OSGi-specific interface that binds the otherwise regular Java classes to the OSGi framework, which manages the life cycle of the bundle.

Each sensor or actuator is represented in the Atlas middleware as an individual OSGi service bundle. Applications and services are also written as

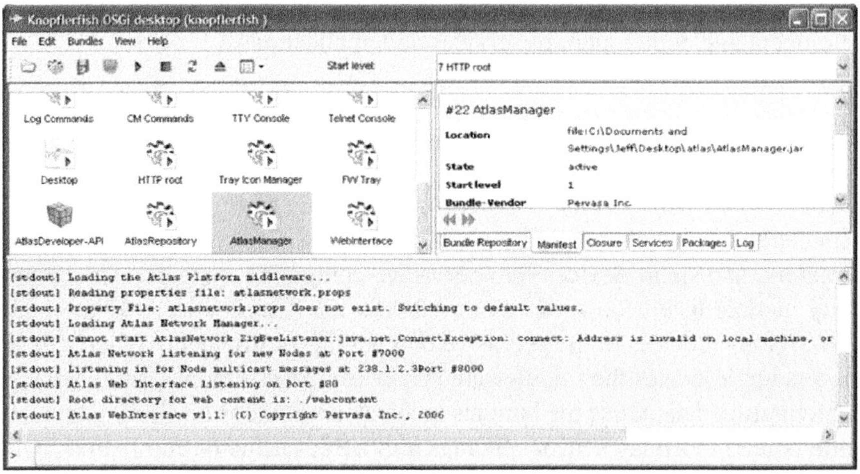

Figure 13. OSGi console with Atlas services running.

bundles. The life cycle management and the discovery service capabilities allow the middleware to manage dynamic environments where devices can come and go, administer dependencies between bundles, enable the composition and activation of more complex applications and services, and allow applications to find and use other existing services. Unlike other discovery services such as JINI and UPnP, OSGi provides a single, centralized runtime environment.

In the Atlas middleware, the network manager bundle handles the arrival and departure of nodes in the network. The configuration manager manages the configuration settings of each node and enables remote configuration. The bundle repository stores and manages all the supported sensor and actuator bundles. Features of these three common services are accessible to the user through an intuitive web interface known as the Atlas Web Configuration and Administration Tool.

In addition to these core bundles, the services layer also contains the virtual sensors framework (Bose et al., 2007), communication and safety modules. Virtual sensors provides support for the reliability enhancement and automatic service composability. A suite of communication modules provide a variety of interfaces, such as web services and a LabVIEW proxy. The context manager employs standard ontology to build a context graph that represents all possible states of interest in a smart space and serves as a safety monitor which ensures that the smart space avoids transition into impermissible contexts. Other safety modules prevent devices and services from performing dangerous operations.

The application layer sits at the top and consists of the execution environment that provides an API to access and control sensors, actuators, and other services. It contains a service authoring tool to enable rapid and efficient development and deployment of services and applications.

3.2 Components of the Atlas Middleware

3.2.1 Connectivity Module.

This module provides a bridge between the physical and digital worlds, allowing easy connection of sensors, actuators, and smart devices into pervasive computing systems. The connectivity module listens on a variety of network interfaces, such as Ethernet and USB, for any devices in the pervasive computing environment. When a device powers up, it locates the middleware server using this module, and exchanges configuration data. Once the bundles associated with the node have started, the node is ready to relay sensor readings and accept actuator commands. Once a sensor or actuator is registered in the Atlas middleware as a service, applications and other services are able to dynamically discover and access them using mechanisms provided by OSGi. The connection between devices and services is maintained by the network manager service.

Network viewer provides a web-based front-end to the network manager, allowing users to view the current network status. It displays the list of active devices and a short summary of their connection statistics and histories.

3.2.2 Configuration Manager.

The configuration manager encapsulates all the methods for recording and manipulating device settings. When a new device uploads its configuration file, the network manager passes it on to the configuration manager, which then parses the file and accesses the bundle repository to get the references for service bundles required by the connected devices. It then loads these service bundles into the OSGi framework, thereby registering the different device services associated with the node. The Configuration Manager web interface (Figure 14) allows a user to view and modify device configurations through the web interface.

3.2.3 Bundle Repository.

The bundle repository manages the various device service bundles required by physical sensors and actuators deployed in the sensor network. The bundle repository eliminates the need for devices to locally store their drivers. Instead, the configuration manager retrieves references to the bundles from the bundle repository when new devices join.

A web-based front-end allow users to view and modify lists of the available service bundles, the physical devices they represent, and other details such as the version numbers and dates uploaded. Users are able to add, delete, and update service bundles, and synchronize with other repositories.

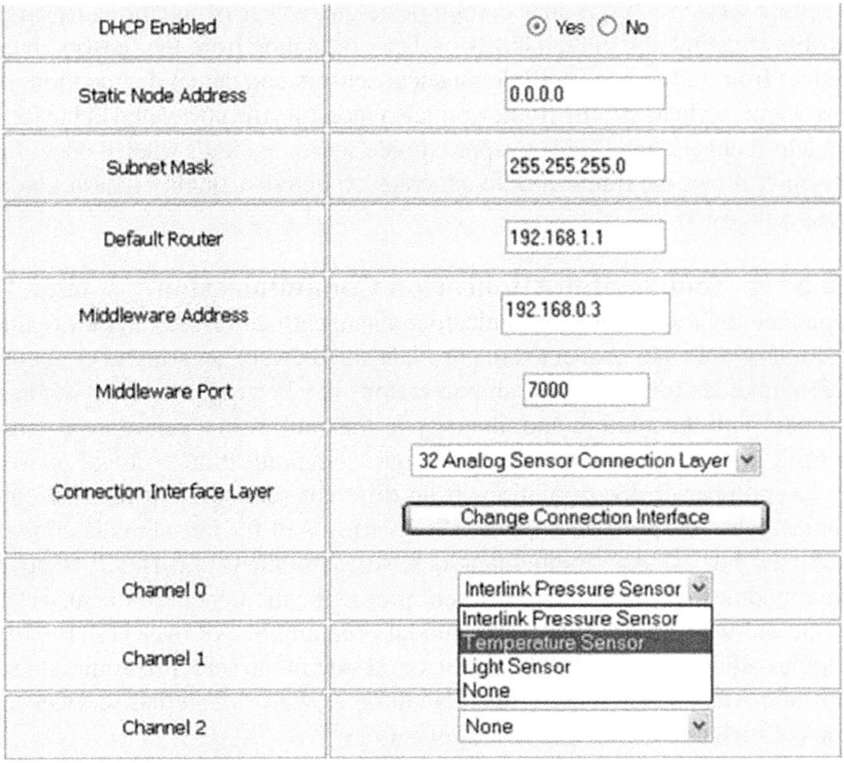

Figure 14. Configuration manager web interface.

3.2.4 Data Processing Modules. Simply retrieving data from each connected sensor and device is neither scalable nor reliable in a normal setting for pervasive computing. Two data processing modules are implemented to address these issues: the on-board processing module installs a digital filter on lower level devices in the network to aggregate data, reduce noise, and allows applications to delegate some decision-making.

The virtual sensors module allows continued operation amid device failures by providing a compensation mechanism and data quality indicator. A virtual sensor is a software service consisting of a group of sensors annotated with knowledge enabling it to provide services beyond the capabilities of any individual component. Virtual sensors may be composed of a group of physical sensors or other virtual sensors, and are classified into three types: Singleton virtual sensor, basic virtual sensor, and derived virtual sensor. Singleton virtual sensor represents a single physical sensor, a basic virtual sensor is composed of a group of singleton virtual sensors of the same type, and a derived virtual sensor is composed of a group of basic and/or other derived virtual sensors of heterogeneous types. Virtual sensor enhances the reliability and availability of sensor services. It provides certain basic guarantees of functionality and is capable of estimating the reliability of data originating from the sensors. It can recover from failures of multiple physical sensors and detect degradation in a sensor's performance. The framework also monitors the correlated behavior of singleton sensors, and is able to approximate sensor readings when devices fail. This also allows the framework to generate explicit data quality measurements for specific services.

3.2.5 Tooling and Application Communication. Instead of implementing a singular communication manager that oversees all the communications within and connected to our Atlas middleware, an array of communication modules for effective communications in a heterogeneous environment is used. With the number and diversity of the entities in a pervasive environment, it is unrealistic to expect one single communication protocol to work on vast number of diverse entities from different vendors. For internal communications, services can use the OSGi wiring API for inter-bundle communications. For external communications, Atlas middleware currently supports three modules realizing three different protocols, the telnet/ssh client, HTTP client, and web service interfacing module running SOAP over HTTP. These modules allow the services and entities to reside in the server to communicate with non-Atlas-based sensors and actuators, as well as external services and systems such as existing business applications.

To support a more flexible and open communication interface while reducing the cost of maintaining the array of communication modules, we have joined forces with several industrial partners in defining an open standard known as *service-oriented device architecture*, or SODA (SODA, 2007). SODA fully embraces service-oriented architecture (SOA) and defines a standard protocol allowing devices and backend systems to communicate in a unified language (de Deugd et al., 2006). Adhering to this standard minimizes the incurred communication cost. A SODA-based communication module is currently under testing, which will serve as the main channel and facilitator for all communications. The existing array of modules will be retained to support legacy systems.

3.3 Programming Support

Creating a pervasive computing space is extremely difficult due to the massive hardware integration task and the distributed nature of the system. It is all too common for specific applications to be "hardwired" to a particular environment. Our goal is to eliminate this difficultly, and two modules are included in the Atlas middleware to support programmability. The Atlas developer API provides standardized interfaces and classes to deal with different devices and applications. The service authoring interface module enables remote service authoring and maintenance, which greatly improves programmers' productivity, and links the Atlas middleware with an IDE, forming an end-to-end solution for service authoring in pervasive computing.

3.3.1 Atlas Developer API. The Atlas developer API provides interfaces and base classes for third parties to develop device and application service bundles on top of Atlas. Programmers wishing to write their own service bundles for sensors or actuators are required to implement the AtlasService interface provided by the API. This interface defines low-level functionality common to all the service bundles, providing a homogeneous interface to heterogeneous devices. The developer API promotes the proliferation of device service bundles. Combined with the bundle repository, it encourages community-based development that can cover large territories of new sensors and actuators.

Using the AtlasClient interface to develop pervasive applications promotes standardized, streamlined interactions between the application and the middleware. Its unified interface allows for rapid development of complex applications over a large set of widely diverse devices.

3.4 The Atlas Platform

The Atlas middleware provides many of the services necessary to create programmable pervasive computing spaces. However, there is still a piece needed to physically connect the numerous and heterogeneous sensors, actuators, and devices – critical pieces of a smart space – with the computing system that hosts the middleware and applications. As shown in the Atlas architecture diagram (Figure 12), this need is addressed by the platform layer.

The Atlas sensor network platform (King et al., 2006) is a combination of hardware and firmware. Together these components allow virtually any kind of sensor, actuator, or other device to be integrated into a network of devices, all of which can be queried or controlled through an interface specific to that device, and facilitates the development of applications that use the devices.

Each Atlas node is a modular hardware device composed of stackable, swappable layers, with each layer providing specific functionality. The modular design and easy, reliable quick-connect system allow users to change node configurations on the fly.

A basic Atlas node configuration (Figure 15) consists of three layers: the processing layer, the communication layer, and the device connection layer.

Figure 15. Three-layer Atlas node.

3.4.1 Processing Layer. The processing layer is responsible for the main operation of the Atlas node. Our design is based around the Atmel ATmega128L microcontroller. The ATmega128L is an 8 MHz chip that includes 128 kB flash memory, 4 kB SRAM, 4 kB EEPROM, and an 8-channel 10-bit A/D-converter. The microcontroller can operate at a core voltage between 2.7 and 5.5 V. We chose this chip for its low-power consumption, plethora of I/O pins, ample SRAM and program space, and the readily available tools and information resources. In addition to the ATmega128L, we include 64 kB of expanded RAM for on-node services and a real-time clock for accurate timing. This clock can also be used to have the microcontroller wake from a sleep state at specified intervals.

Whereas most current commercially available platforms are concerned only with sensors. Atlas treats actuators as a first-class entity, and the power supply design on the processing layer reflects this design: the platform supports both battery power and wired power.

Battery power is an obvious choice in sensor networks, where applications often require long-lived, unattended, low-duty cycle sensing. The other option, wired power, may seem unusual in this field. But given the primary goal of Atlas – enabling the development of smart spaces – wired power is a necessity. Smart spaces will contain many actuators, and driving these devices requires much more power than operating the platform nodes. Hence, Atlas supports both wired and battery power.

In case wired power is used, the second plug can be used to daisy-chain nodes together, reducing the number of outlets used in a smart house environment. The number of nodes that can be chained to a single power supply depends on the number and type of devices connected to the platforms. For example, in the Gator Tech Smart House smart floor, each node is connected to 32 pressure sensors and 15 Atlas nodes can be daisy chained easily.

3.4.2 Communication Layer. For a sensor and actuator network platform to be useful, users must be able to access the data being produced by sensors, and must be able to send commands to the actuators. With Atlas, data transfer over a network is handled by the communication layer. Several options are currently available:

- Wired 10/100 Base-T Ethernet;
- IEEE 802.11b WiFi;
- ZigBee;
- USB.

Wired Ethernet. Wired Ethernet is important in situations requiring high-speed data access over an extremely reliable connection. For example, the Gator Tech Smart House uses Wired Ethernet Atlas nodes for critical systems such as location/fall detection. It is ideal for applications where nodes are situated in areas shielded from RF communication, as in many industrial settings, or for deployments where jamming from benign or malicious external signals may be an issue. Wired Ethernet is also preferable in high-security settings where snooping must be detected, as splicing the Ethernet cable produces a change in the impedance of the wires that can be sensed.

The current Atlas wired Ethernet communication layer uses the LANTRONIX XPort. It is an integrated Ethernet device, meaning the module includes its own microcontroller, which operates the Ethernet transceiver and runs a full TCP/IP stack. The XPort provides 10/100 Mb networking, and includes higher level protocols such as DHCP, HTTP, and SMTP.

WiFi. The WiFi communication layer is based on the DPAC WLNB-AN-DP101 airborne wireless LAN module, providing 802.11b connectivity to the Atlas platform. Like the XPort, the DPAC module is an integrated device, with its own microcontroller to operate the device and implement the network protocols.

Also like the Wired Ethernet layer, the WiFi layer, which provides connection speeds up to 11 Mbit, is appropriate for situations requiring high-speed data access. The 802.11b devices are typically rated for a range of 50 m, though the exact range depends on antennas used and environmental effects.

WiFi is not a low-power standard. The WiFi communication layer is best used when wired power is available but wired Ethernet is not. Battery operation is possible, but an extended life is possible only with very infrequent transmissions.

Figure 16. Atlas ZigBee node in ruggedized packaged form factor.

ZigBee. The ZigBee communication layer (Figure 16) uses the Cirronet ZigBee module. This module is based on the IEEE 802.15.4 standard for low-power wireless networking. Atlas nodes using ZigBee communication are the best choice for untethered, extended-life, battery-operated applications. ZigBee-based Atlas nodes can function exactly like other Atlas nodes by means of the Cirronet ZigBee Gateway, or can form an ad hoc mesh network for non-pervasive-computing sensor network scenarios.

USB. The USB communication layer allows Atlas nodes to connect directly to the middleware computer using its USB ports. The USB layer is primarily used for secure programming and configuration of nodes – because the node is connected directly to the computer, information such as secret keys can be passed to the node without fear of the data being compromised.

3.4.3 Device Interface Layer. The interface layer is used to connect the various sensors and actuators to the Atlas platform. Interface layers are available for a variety of analog and digital sensors, actuators, and complex third-party devices.

Analog Sensors. Two-device interface layers are available for analog sensors. Each board accepts standard 3-wire analog sensors. The first interface layer, the 8-sensor board, supports up to eight analog sensors, with polarized plugs to ensure the proper orientation of any device connected. The second analog sensor interface layer, the 32-sensor board supports 32 sensors. This allows for very cost-effective deployments, but due to space limitations, this board does not include polarized headers, so users must be careful to plug sensors in using the correct orientation. The reference, ground, and signal pin rows are labeled on the board to help users correctly connect their devices.

Digital Sensors and Actuators. The digital contact interface layer (Figures 4–10) supports up to 16 contact or other two-pin digital sensors or actuators. Since two-wire digital devices can be plugged into the layer in either orientation, this board does not include polarized headers.

Servos. The servo interface layer allows six servo motors to be controlled by the Atlas platform using pulse width modulation (PWM). The servo interface layer also includes the dual-head power connectors. Most servos require at least 7 V, more than the ATmega128 L can provide. Servos can either use a pass-through cable from the processing layer, or can be directly connected to a separate DC power supply.

General Purpose I/O. The general purpose I/O interface layer (Figures 4–12) allows users to connect any device to the Atlas platform without requiring a customized interface layer. It also allows for a mix of analog and digital sensors and actuators to be controlled by a single Atlas node. This is accomplished with a pair of standard RS232 serial port connectors, allowing any serial device to be integrated into an Atlas deployment.

3.4.4 Atlas Node Firmware. The firmware runs on the processing layer of the Atlas platform hardware, and allows the various sensors, actuators, and the platform itself to automatically integrate into the middleware framework. Unlike other sensor network platforms, application developers do not work on the firmware level. The Atlas firmware is a fixed system specifically designed to integrate the physical node and any connected devices with a server running the Atlas middleware.

As a modular platform, at boot the Atlas firmware first detects the type of communication layer attached to the node. This allows infrastructure to change or nodes to be moved without requiring any modifications to running applications. The firmware continues to initialize the node, and then seeks out the middleware server (Figure 17).

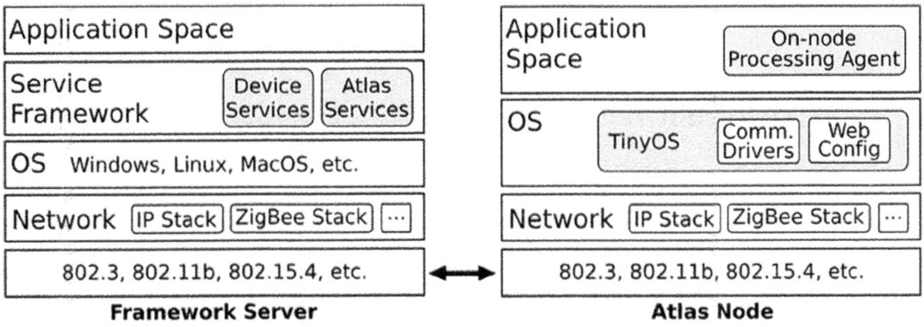

Figure 17. Software architecture of the Atlas platform.

After connecting to the server, the node is able to integrate with the middleware. As mentioned, the connection module of the middleware allows the node to upload its configuration details, which registers all the devices connected to that particular node, making the devices available as OSGi services in the framework.

After this initialization process concludes, the node goes into data-processing mode. It begins sending data from sensors and receiving commands to control sensor and actuator operations.

4. Status of the Gator Tech Smart House

The Gator Tech Smart House has been in operation for nearly 3 years, since its grand opening on January 28, 2005. Reliability has been tested through frequent tours, and the usefulness of its offered services has been demonstrated through numerous study groups. Live-in trials began in the house on March 24, 2006. The subjects' activities were monitored and logged for analysis both by our team and by collaborators. New services and systems have been introduced periodically to the Gator Tech Smart House, and the space has seen two major upgrades of infrastructure technology, such as the expansion of the smart floor from just the kitchen area to the entire house (quintupling the number of sensor platforms and force sensors deployed for that one project), with minimum downtime.

In addition to services targeted at independent living for senior persons, the Gator Tech Smart House is currently being expanded to offer health-care-oriented services. Through funding from the National Institutes of Health (NIH) and consistent with NIH's Roadmap for Medical Research, we are repurposing the Gator Tech Smart House as a smart space supportive of the obese and the diabetic. While some services in the house will carry over, such as those designed to assist mobility-impaired residents, the majority of new applications will now cover three primary concerns: dietary monitoring, quantification of activity level, and assistance in vital-sign testing (insulin level, blood pressure, etc.).

5. Conclusion

This repurposing is a powerful test of our Atlas platform for programmable pervasive computing spaces. Many of the systems deployed in the house, such as the SmartWave and the smart floor, will be used in the new applications. Since these physical systems are automatically translated into software services, integrating them into the new applications is trivial. Additionally, the plug-and-play development model offered by Atlas allows us to bring in new devices, such as digital scales and blood pressure monitors, without requiring tedious system integration during the move from lab to house.

We are also refining our enabling Atlas platform. Improvements to the sensor network platform, middleware, and associated tooling will continue to facilitate the development process. Other projects in the lab will expand the self-sensing spaces concept, further reducing the role of engineers in deploying a working pervasive computing environment. Ultimately, our goal is to create a "smart house in a box": off-the-shelf assistive technology for the home that the average user can buy, install, and use.

References

Bose, R., Helal, A., Sivakumar, V., and Lim, S. (2007). Virtual Sensors for Service Oriented Intelligent Environments. In *Proceedings of the 3rd IASTED International Conference on Advances in Computer Science and Technology*.

de Deugd, S., Carroll, R., Kelley, K., Millett, B., and Ricker, J. (2006). SODA: Service-Oriented Device Architecture. *IEEE Pervasive Computing*, 5(3)94.

El-Zabadani, H., Helal, A., Abdulrazak, B., and Jansen, E. (2005). Self-Sensing Spaces: Smart Plugs for Smart Environments. In *Proceedings of the 3rd International Conference On Smart Homes and Health Telematic (ICOST2005)*.

El-Zabadani, H., Helal, A., and Schmalz, M. (2006). PerVision: An Integrated Pervasive Computing/Computer Vision Approach to Tracking Objects in a Self-Sensing Space. In *Proceedings of the 4th International Conference On Smart Homes and Health Telematic (ICOST2006)*.

GTSH (2007). The Gator Tech Smart House Web Site (http://www.icta.ufl.-edu/gt.htm).

Helal, A. (2005). Programmable Pervasive Spaces. *IEEE Pervasive Computing*, 4(1):84–87.

Helal, A. and Mann, B. (2003). Integrating Smart Phones with Smart Spaces for Elder Care. In *Proceedings of the 1st International Conference On Aging, Disability and Independence*.

Helal, A., Mann, W., Elzabadani, H., King, J., Kaddourah, Y., and Jansen, E. (2005). Gator Tech Smart House: A Programmable Pervasive Space. *IEEE Computer*, 38(3):50–60.

Kaddourah, Y., King, J., and Helal, A. (2005). Cost-Precision Tradeoffs in Unencumbered Floor-Based Indoor Location Tracking. In *Proceedings of the 3rd International Conference On Smart homes and health Telematic (ICOST2005)*.

King, J., Bose, R., Zabadani, H., Yang, H., and Helal, A. (2006). Atlas: A Service-Oriented Sensor Platform. In *Proceedings of the 31st IEEE Conference on Local Computer Networks (LCN)*.

Lee, C., Nordstedt, D., and Helal, A. (2003). OSGi for Pervasive Computing. *IEEE Pervasive Computing*, 2(3):89–94.

Maples, D. and Kriends, P. (2001). The Open Services Gateway Initiative: An Introductory Overview. *IEEE Communications*, 39(12):110–114.

Russo, J., Sukojo, A., Helal, A., and Davenport, R. (2004). SmartWave Intelligent Meal Preparation System to Help Older People Live Independently. In *Proceedings of the 2nd International Conference on Smart Homes and Health Telematic (ICOST2004)*.

SODA (2007). SODA Standard Web Site (http://www.sensorplatform.-org/soda).

Chapter 2

DO DIGITAL HOMES DREAM OF ELECTRIC FAMILIES? CONSUMER EXPERIENCE ARCHITECTURE AS A FRAMEWORK FOR DESIGN

Brian David Johnson
The Intel Corporation
brian.david.johnson@intel.com

Abstract If we are designing for digital homes then we are not designing for humans? How do we truly design for real people? Consumer experience architecture (CEA) provides an actionable framework for the development, design, and production of products and services specifically centered around human needs, desires and frames of understanding. This chapter dismantles CEA into its essential components, exploring real-world examples and illustrations. Finally the chapter challenges the reader to expand current development practices by looking toward science fiction or other cultural inputs as possible laboratories or inspirations for future designs.

Keywords: Product innovation; Holistic design framework; Human–computer interaction; Ethnography in design; Designing for humans; Science fiction as laboratory.

1. Introduction

The title of this chapter takes its inspiration from the title of Philip K Dick's 1968 science fiction masterpiece "Do Androids Dream of Electric Sheep?" The novel tells of the moral conflict of Rick Deckard, a bounty hunter who tracks down androids in a devastated futuristic San Francisco. The novel was popularized in the early 1980s when Ridley Scott directed the film Blade Runner,

W. Minker et al. (eds.), *Advanced Intelligent Environments*,
DOI 10.1007/978-0-387-76485-6_2, © Springer Science+Business Media, LLC 2009

based loosely on Dick's story. One of the most enduring themes of the book is what it means to be human and conversely what it means not to be.

I wanted to make reference to Dick's novel because I am interested in what it means to design for humans. How do you develop and design future technologies for people? What makes these new products valuable? What makes them usable and meaningful? Similarly, what happens when you design without humans in mind? What happens when products are designed without an understanding of the people who are going to use them? When we design digital home products are we designing them for electric families instead of real people? Does anyone really want a digital home or do they just want their existing homes to be just a little bit better? In this chapter I explore consumer experience architecture as a practice and a methodology for developing products and services so that they fit intuitively into the lives of consumers. Here, I draw on recent experiences at Intel Corporation, where we have applied this framework directly to the development of personal technology devices.

Consumer experience architecture (CEA) provides a framework for multiple inputs into the design and development process, including ethnographic research, market analysis, demographic profiles, competitive analysis along with technological innovation and exploration. CEA provides a holistic framework that can be used by technology and social science researchers, product planners, hardware and software engineers as well as project managers to unite their varied domains into a process that holds the human value of the product as the guiding principle throughout that product's development.

Additionally, CEA provides the ability to identify, specify, document, and validate the human value of the product as the desired consumer experience. By documenting this experience, founded on both the human insights and technological innovation, it can then be systematically validated at key milestones in the development process. This rigorous documentation and validation of the consumer experience means that we can develop products that can be both futuristic and fit effortlessly into people's daily lives.

Finally, once we have implemented CEA's holistic approach to technology development, we are free to ask ourselves what other influences could be utilized in the design of new products. An interesting and entertaining challenge would be to take the work of Philip K. Dick and other science fiction writers, using their visions of the future as another input into the CEA process. There has always been a close tie between science fiction and science fact. Could the CEA framework create wildly futuristic devices that would still have meaning and value for consumers?

2. User Experience Group Overview: Understanding People to Build Better Technology

In 2005, Intel underwent a significant restructuring which included the establishment of several new business groups focused explicitly around usage of ecosystems and activities – the home, the office, emerging markets, and mobile users. As part of the restructuring, senior executives also endorsed the inclusion of user research teams and competencies. In the newly established Digital Home Business Group, an explicit commitment to consumer-centric thinking has been an important part of business from day 1. The User Experience Group, of which I am a member, is an interdisciplinary team dedicated to bringing the perspectives of ordinary people into Intel's product planning, development, and marketing activities. For the last 2 years, I have been a consumer experience architect within this group.

Our group includes two distinct competencies: one with quantitative and qualitative research focus and the other oriented more closely to usability, usage modeling, and user experience assessment. Our research competency, which consists of social science and design researchers, spends time in people's homes all over the world. We take as a starting point the firm conviction that people's social and cultural practices change far more slowly than technologies. This team is focused on getting a sense of what makes people tick, what they care about, what they aspire to, and what frustrates them. This research is focused around getting a sense of the larger cultural patterns and practices that shape people's relationships to and uses of new technologies.

In 2006, we conducted more than 400 field interviews in 16 countries, and the team is on track for similar metrics in 2007. To accomplish this research we use long-standing qualitative and interpretive studies such as participant observation, interviews, as well as shadowing people's daily lives. Typically these are on small scale, conducted in person by the team, and are based on a traditional approach of ethnographic field research (Salvador et al., 1999). Along with this we will also use more experimental design research methods such as cultural probes, photo diaries, cognitive mapping, and story telling exercises (Gaver et al., 1999). These contemporary methods are a means to involve the participants in a more collaborative way during the research. Often we send design research activities to the participants before the research team arrives. This prompts the participant to begin documenting their lives right away and provides us a rich starting place to begin the ethnographic research.

3. Guiding Principles for Global Research and Product Investigation

Our research activities are guided by three principles: privileging people, practices, and presence. First we focus on people not users. It can be an unfortunate trap for many product development teams to conceptualize the people who will be buying and/or using their product as simply a user of that specific product. They do not envision or comprehend the wider life and influence on their customer. This conceptualization does not see them or treat them like a human; much like this chapter's title it treats the user more like a digital family than a flesh and blood user. The result of these digital fantasies can be quite shocking and are rendered most visible when the person who is looking to buy or use the product does not know how to use it. On some occasions, the consumer may never understand the value of the product and simply ignore it.

Our second guiding principle concerns social and cultural practices: we are interested in people's everyday lives. We look for domesticated technologies as opposed to imagined technologies. Much like design teams conceptualize people as simply users or non-humans, these same computer science or development teams can imagine their technologies as theoretical or engineering prototypes. What is lost in this approach is that all technologies exist in the real world once they have left the lab. And we all know the real world is a very different place than the lab. Because of this, when we explore how people all over the world are using technology, we make sure to look at how that technology is actually used. What do people do with this technology in their lives? What works for them? What does not work? How and why does the technology break down? Who spends time using the device or service? In this way we begin to form a grounded and realistic vision of how technologies are used by people everyday.

Our third and final guiding principle is that we always make sure to keep in mind that most of people's lives are spent off-screen, meaning that most people's lives are spent not sitting in front of a screen or even using technology. In fact this off-screen time is the time that most people cherish most. To understand this life off-screen and understand why it fuels people, we explore the meaning people get from every aspect of their lives.

4. Houses are Hairy: The Need for Experience Design

"Experience Design has become newly recognized and named. However, it is really a combination of many previous disciplines; but never before have these disciplines been so interrelated, nor have the possibilities for integrating them into whole solutions been so great" (Shedroff, 2001).

A few years ago I was building a personal computer with a friend of mine. He is a software engineer for a near-by science museum. We were talking about where to put the computer once it was built. My question was, "Should I put it on the floor of the study or on a table?" He said it really did not matter. "But there's so much more dust and dirt on the floor. It has to matter," I replied. "Brian, you have no idea how much dust and hair there is in your house... In everyone's house," he replied. "If a hardware engineer ever opened up my computer or any appliance in my house they would be shocked and horrified with what they found. Our houses aren't clean rooms. What can you do? Houses are hairy; it doesn't matter where you put it."

My friend made a good point; houses are "hairy" and many products, especially technology products like computers, are not always designed for the cluttered lives of humans (Figure 1). But this example goes far beyond the physical. It can be argued that the physical designs of products are actually far more suited to consumers than their wider needs for purchase, set up, maintenance, and ongoing use. Not only houses are hairy but humans lives are also busy and wonderfully cluttered with a vast array of influences that affect how they understand and use technology. In short, the entire consumer experience of many products appears not to be designed with the real life of their consumers in mind.

Figure 1. The PC goes home.

"Whereas architecture and furniture design have successfully operated in the realm of cultural speculation for some time, product design's strong ties to the marketplace have left little room for speculation on the cultural function of electronic products. As ever more of our everyday social and cultural experiences are mediated by electronic products, designers need to develop ways of exploring how this electronic mediation might enrich people's everyday lives" (Dunne, 2006)

Consumer experience architecture, as it can be applied as a framework for the research, design, development, and marketing of technology, is a powerful tool. It allows companies like Intel to hardwire the real lives and desires of humans into a process it too often oriented more toward an engineering culture. With the increasing complexity of digital home products and services, understanding and architecting consumer experiences is becoming more important and essential for success.

5. Consumer Experience Architecture in Industry

"Consumer experience will drive the adoption of home media technology, not a particular piece of equipment" (Kim, 2007).

At Intel and across the high-technology development industry, CEA, or more specifically the desired result of consumers' acceptance of new devices and services is gaining exposure and relevance. This increased exposure and acceptance has everything to do with financial success. A recent Parks and Associates Digital Home Services Report (2007) found that as many as 40% of people purchasing wireless networking equipment to connect computers and other devices in their homes return them to the store for a refund. The alarming part of this statistic is that of the 40% that were returned, 90% of these devices had no known defect when the returned merchandise was checked. From this information one can extrapolate that people were returning the devices because they did not understand them, did not value them, or simply could not make them work. This is just one example of many. The rising complexity of devices in the market means that this problem will only continue unless there is a significant cultural shift in the way that devices and products are developed for the general public.

Companies are seeing that even if their devices are innovative and priced right consumers may still not buy them if they do not understand how to set them up and use them. Worse yet, people will return products if their experience with the product does not match what they thought they were buying or what the manufacturer had promised.

6. Technology for Humans: A Design Framework

CEA provides a framework that we can use to insert the consumer's perspective at key points in the product development process. At Intel, the cycles of planning and development (prototype to alpha and beta and release candidates) are intersected at key points to ensure that the original goals of the product and the value of the product to the general public are always met (see Figure 2).

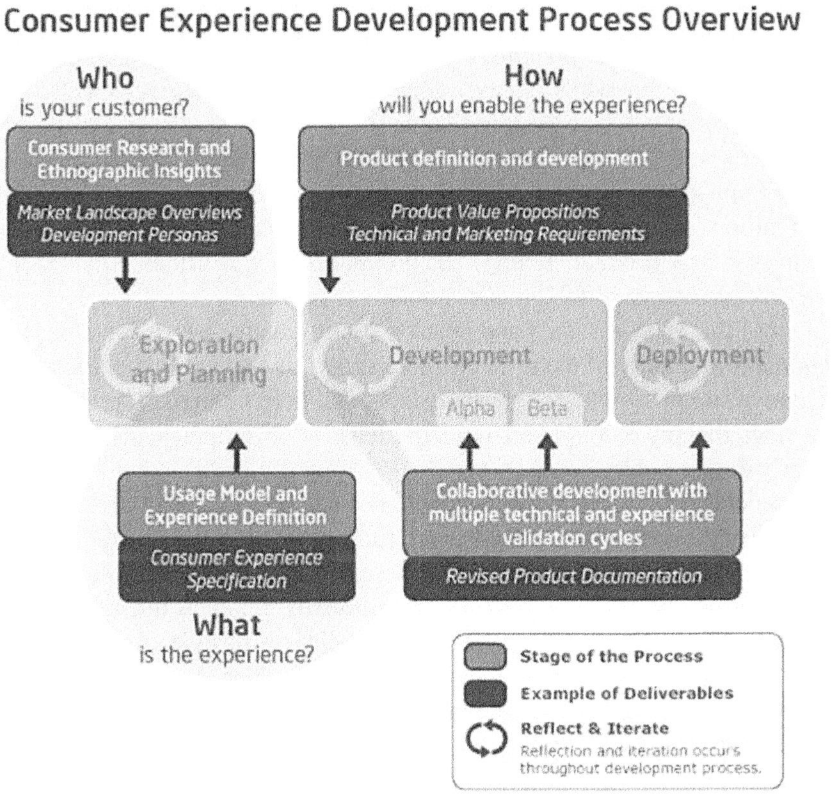

Figure 2. Overview of development process.

This development process can be broken up into four discrete and distinct stages. Each stage serves as a key point of intersection, influence, and iteration in the development process.

6.1 Stage 1: Human Insight

I was once asked, if by using the CEA process, could I have predicted the important and massive public acceptance of e-mail. I replied, that yes, I probably could have recognized the significance of e-mail as a technology in which people would be wildly interested. The reasoning was simple: for hundreds of years people in a range of different social, cultural, and economic circumstances had been composing, writing, and sending letters to one another, and of course, for thousands of years before that, oral messages conveyed thoughts, emotions, and information over distances, small and great. It is at the foundation of how we communicate with our friends and family. E-mail was simply a new means of distribution for a very old and cherished form of social interaction – communication.

The initial research and information gathering stage of the CEA framework provides input into the planning cycle. Here the team's ethnographic insights are coupled with market strategy, competitive product analysis as well as technical innovations. It is important to note that for many new products there may be little to no existing competitive or market information. In this case, ethnographic and other forms of qualitative consumer-centric information become even more valuable, as it provides a foundation of human values and behavior around the new product. Even if the product is new, the human behavior that will take advantage of the product remains the same.

Out of this early research and persona development, key deliverables are an actionable summary of the research. Typically this includes a top line report or executive summary with appropriate detail and field findings. It is important at this stage that the recommendations or areas for development serve as a guide in early design cycles.

A second deliverable from this cycle is a set of personas or archetypes that describe the people for whom the product is being designed (the "Who" in Figure 2). Utilizing personas in the design and development of products is not a new practice (Cooper, 2004). Traditionally personas utilize market and demographic information to create a personality or lifestyle. A way to expand this sometimes-limited approach can be the addition of real-world data and insights. Archetypes, as they are sometimes called, can consist of a collection of ethnographic family and participant profiles that outline actual people that have been observed and studied. The collection of these profiles combined with demographic and market information can provide a more in-depth portrait of the target consumers with a depth that is grounded in actual human interactions.

6.2 Stage 2: Experience Definition

"New cognitive models can often revolutionize an audience's understanding of data, information, or an experience by helping them understand and reorganize things they previously understood (or, perhaps, couldn't understand), in a way that illuminates the topic or experience" (Shedroff, 2001).

As the planning cycle moves forward and the product becomes more defined, a set of documents are created that outline the specific consumer experience that the product or service is trying to bring to market. A benefit of this stage is that it provides the opportunity for every member of the development team to gain a holistic understanding of the desired consumer experience. From the technical developers to marketing team, this knowledge proves to be invaluable as the development cycles move forward. It provides both a base of knowledge from which each team member could draw upon to inform their specific domains in the design process. This knowledge becomes a shared understanding between all team members. It gives them a common language and enhances collaboration. Additionally, it gives them a shared goal that has been documented and can be retuned to for wider problem-solving activities of even broader corporate or business group alignment. This experience definition can help bridge the process gaps that occur between engineering and marketing or hardware and software teams or even project teams and management.

The experience specification builds upon the early research and persona development and identifies experience opportunities or specific human values that the product can enhance. As stated previously, consumer experience is the sum total of multiple inputs or influences on the consumer understanding of a product. All of these inputs serve to form a mental model for the consumer. It is this mental model that we can use to construct and develop a solid experience that will be both usable and desirable.

Each of these influences can be mapped and explored in an in-depth review of the product's life cycle. This process begins with the consumers' first awareness of the product, typically through advertising or marketing. This can also occur through the consumers' social network of friends and family. From this point the product life cycle documents the consumer's behaviors as they gather more information, research the product, and ultimately use or touch the product for the first time. This first experience can occur in a retail setting or even online. The life cycle then outlines the purchase environment either in a retail store or online and then the physical out of box experience. This step in the process should be specific, recording whether the appropriate documentation and cables are included, whether the printed package design continues to deliver on the product's marketing and brand promise, even if the packing materials are easily recycled. Finally we follow the product through its initial instillation and set-up, ultimately exploring the complexity of the products daily use by multiple consumers in the household.

This exhaustive documentation and visualization affords the development team a framework to envision the product and comprehend the overarching consumer experience at its earliest stage of development. It uncovers details in every step of a complex process that are typically overlooked.

The consumer experience specification becomes a core document in the product's development library, consulted by new team members, reviewed by the team in problem-solving brainstorms, and also as a foundation for the third stage in the framework.

6.3 Stage 3: Early Product Definition

Once the experience opportunities have been identified and the consumer's experience mapped, it is necessary to deconstruct these opportunities into usage models and values propositions. Usage models are an industry-accepted standard format for the development of technology specifications and prototypes. Usage models contain the detail necessary to translate usage information to a set of user requirements to guide planners, architects, and engineers in generating hardware and software requirements. Usage models include the following:

- Usage summaries: A descriptive summary of the usage (e.g., text, storyboards, concept drawings)
- Use cases: A collection of related interactions between users and system (e.g., streaming video content from home PC to mobile phone, co-editing video simultaneously from two PCs in different locations)
- Usage scenarios: Stories or explorations that illustrate how people or the archetypes in a specific context actually use the system to accomplish their goals
- Task flows: A visual representation of the step-by-step course of events required for the usage to occur in a positive way
- Operational profiles: The operations a person can perform with the system along with how frequently each will be performed relative to the others

From the experience opportunities and usage models we then develop the product's value propositions. These value propositions act as an expression of the product to the consumer, using their own language. Documenting these value propositions in consumer-specific language is an essential part of the framework. Many times in the development of products the development team can use their own corporate or engineering-based terms and vocabulary to describe this value. The team uses this language to describe to themselves and their management the benefit of the product to the consumer. This practice opens up a gap between the development team and the people who will ultimately use the product. Not surprising the average person would not understand the corporate and engineering terms used in most development companies. Using this language further separates the production team from the people they are designing for.

Clearly development teams need their engineer cultures to operate as a business but at the same time it is important that they also take a moment and speak the product's value propositions in the language of the personas or archetypes that were defined in the first stage of the process.

This step in the framework serves as a point of reflection and iteration. It allows the team to make minor adjustments to their products personas and minor course corrections in the experience that is being developed. In this way the team can track their progress. Also this articulation can serve as a way to discuss the attributes and value of the product to people both inside and outside the development team. It becomes a kind of shorthand or elevator pitch that can be used to explain the product to management, outside companies, or investors.

Along with this reflection and iteration the product's experience opportunities and value propositions are formalized into usage models. The usage models provide the in-depth detail needed for engineering to develop the product to the point of execution. The details of a full usage model definition should encompass the full specifications of the product. Again the framework provides the team a means to visualize the product down to the smallest detail before they begin building. Here issues of technical feasibility can arise and possible adjustments to the product will need to be made. Likewise, marketing and business teams' involvement can uncover underlying customer feasibility.

6.4 Stage 4: Production and Validation

The final step in the consumer experience framework is the longest in duration and the most complex in execution. During the product development and validation cycle the team applies a user experience (UX) validation process or UX process throughout the entire production of the product. The UX process encompasses a variety of systematic methods employed to evaluate and understand people's perceptions and experiences with the product. UX's targeted methods examine the user experience with concepts, prototypes, functional product, and competitor products. UX is not market research or focus group testing, but rather assessment of people's actual interactions with a prototype or product of some sort.

At each key milestone in the development process (e.g., prototypes, alpha, beta, and release candidates) the team uses UX to validate that the original consumer experience goals are being met by the product. The test protocols for the UX validation are based on the core documents of the consumer experience framework. The archetypes and personas establish the audience for the UX test. The experience specification describes the test environments and how the product should present itself to the consumer. Finally the value propositions can be tested to see if they do indeed have value to the consumer and if the product is meeting the promise of these propositions.

The UX validation process provides iterative feedback directly from the consumer as to the successes and failures of the product. By performing this validation process multiple times throughout development and basing all stages on a consistent framework UX allows the development team to refine the product multiple times to meet the original experience opportunities outlined for the product.

The results of the UX validation process are not only valuable to the development team. The iterative results of this process coupled with the experience documents from previous stages of the framework provide a clear and compelling picture of the product even before it has been shipped. The results of the UX validation can provide clarity to upper management, possible partners as well as the investment community.

7. Conclusion: How I Learned to Stop Worrying About the Future and Love Science Fiction: A Challenge

The CEA framework, as outlined in these four stages, provides a systematic approach to ensure that products are both grounded in human values and that these values are delivered on throughout the development process. From initial research to the final validation, the CEA framework lays a solid foundation upon which all team members can base their specific innovations, assured that their efforts will resonate with the intended audience.

Now that we have established an actionable framework for the application of human-centered values and experience to the product development process, it allows us to examine other inputs we might use in this process.

An interesting challenge would be to examine how we could utilize the undeniable power of futuristic visions exemplified in the inspirational visions of science fiction to act as another meaningful input into the CEA process. Traditionally science fiction would be categorized as a specific cultural influence that acts upon the consumer. But I would argue that science fiction occupies a unique position in the influence of consumers' view of technology. It gives people not only a vision of what these future innovations might be but it also uses story and character to give these innovations a wider context so that they can be better understood and valued.

"It is my contention that some of the most remarkable features of the present historical moment have their roots in a way of thinking that we have learned from science fiction" (Disch, 2000).

Inventors and developers have always been influenced by science fiction. Video phones imagined in fictions like *2001 A Space Odyssey* are available for purchase right off store shelves. What is the latest mobile phone but a realization of the Star Trek communicator? The CEA framework now provides an overt way to incorporate these inventions of science fiction and turn them into science fact.

Acknowledgments This chapter could not have been written without the support of Genevieve Bell and the Intel Corporation. The works of Michael Payne and Cory Booth were an essential component in the CEA framework.

References

Cooper, A. (2004). *The Inmates Are Running the Asylum: Why High Tech Products Drive Us Crazy and How to Restore the Sanity.* Sams-Pearson Education.

Disch, T. M. (2000). *The Dreams Our Stuff is Made of: How Science Fiction Conquered the World.* Free Press.

Dunne, A. (2006). *Hertzian Tales.* The MIT Press.

Gaver, W., Dunne, T., and Pacenti, E. (1999). Cultural Probes. *Interactions,* 6(1):21–29.

Kim, E. (2007). *As Quoted in Digital Home: It's the Experience Not the Device.* Schwankett, Steven, PC World.

Salvador, T., Bell, G., and Anderson, K. (1999). Design Ethnography. *Design Management Journal,* 10(4):9–12.

Shedroff, N. (2001). *Experience Design.* Waite Group Press.

Chapter 3

AN ARCHITECTURE THAT SUPPORTS TASK-CENTERED ADAPTATION IN INTELLIGENT ENVIRONMENTS

Achilles D. Kameas
DAISy group, The Computer Technology Institute, University of Patras Campus
Patras, Hellas
School of Sciences and Technology, The Hellenic Open University,
Patras, Hellas
kameas@cti.gr

Christos Goumopoulos
DAISy group, The Computer Technology Institute, University of Patras Campus
Patras, Hellas
goumop@cti.gr

Hani Hagras, Victor Callaghan
The Computational Intelligence Centre, Department of Computing and Electronic Systems, University of Essex, Wivenhoe Park, Colchester, UK
{hani,vic}@essex.ac.uk

Tobias Heinroth
Institute of Information Technology, Ulm University, Ulm, Germany
tobias.heinroth@uni-ulm.de

Michael Weber
Institute of Media Informatics, Ulm University, Ulm, Germany
michael.weber@uni-ulm.de

W. Minker et al. (eds.), *Advanced Intelligent Environments*,
DOI 10.1007/978-0-387-76485-6_3, © Springer Science+Business Media, LLC 2009

Abstract The realization of the vision of ambient intelligence requires developments both at infrastructure and application levels. As a consequence of the former, physical spaces are turned into intelligent AmI environments, which offer not only services such as sensing, digital storage, computing, and networking but also optimization, data fusion, and adaptation. However, despite the large capabilities of AmI environments, people's interaction with their environment will not cease to be goal-oriented and task-centric. In this chapter, we use the notions of ambient ecology to describe the resources of an AmI environment and activity spheres to describe the specific ambient ecology resources, data and knowledge required to support a user in realizing a specific goal. In order to achieve task-based collaboration among the heterogeneous members of an ambient ecology, first one has to deal with this heterogeneity, while at the same time achieving independence between a task description and its respective realization within a specific AmI environment. Successful execution of tasks depends on the quality of interactions among artifacts and among people and artifacts, as well as on the efficiency of adaptation mechanisms. The formation of a system that realizes adaptive activity spheres is supported by a service-oriented architecture, which uses intelligent agents to support adaptive planning, task realization and enhanced human–machine interaction, ontologies to represent knowledge and ontology alignment mechanisms to achieve adaptation and device independence. The proposed system supports adaptation at different levels, such as the changing configuration of the ambient ecology, the realization of the same activity sphere in different AmI environments, the realization of tasks in different contexts, and the interaction between the system and the user.

Keywords: Ambient intelligence; Pervasive adaptation; System architecture; Ambient ecology; Activity sphere; Ontology; Ontology alignment; Agents; Fuzzy agents; Interaction; Interaction modality; Pro-active dialogue.

1. Introduction

Ambient intelligence (AmI) is a new paradigm that puts forward the criteria for the design of the next generation of intelligent environments (Remagnino and Foresti, 2005). The realization of the vision of ambient intelligence requires developments both at infrastructure and application levels. As a consequence of the former, physical spaces are turned into intelligent environments, which offer not only services such as sensing, digital storage, computing, and networking, but also optimization, data fusion, and adaptation. Intelligent computation will be invisibly embedded into our everyday environments through a pervasive transparent infrastructure (consisting of a multitude of sensors, actuators, processors, and networks) which is capable of recognizing, responding, and adapting to individuals in a seamless and unobtrusive way (Ducatel et al., 2001). Such a system should also provide the intelligent "presence" as it be able to recognize the users and can autonomously program itself in a non-intrusive manner to satisfy their needs and preferences

(Doctor et al., 2005). AmI offers great opportunities for an enormous number of applications in domains such as health care, the efficient use of energy resources, public buildings, and in leisure and entertainment. Ubiquitous computing applications constitute orchestrations of services offered both by the environment and the information devices therein (Kameas et al., 2003).

Every new technological paradigm is manifested with the "objects" that realize it. In the case of AmI, these may be physical or digital artifacts. The former, also known as information devices, are new or improved versions of existing physical objects, which embed information and communication technology (ICT) components (i.e., sensors, actuators, processor, memory, wireless communication modules) and can receive, store, process, and transmit information. The latter are software applications that run on computers or computationally enabled devices (i.e., digital clocks, MP3 players, weather forecasts etc). Thus, in the forthcoming AmI environments, artifacts will have a dual self: they are objects with physical properties and they have a digital counterpart accessible through a network (Kameas et al., 2005). We shall use the term ambient ecology to refer to a collection of such artifacts that can collaborate to achieve a given task.

An important characteristic of AmI environments is the merging of physical and digital space (i.e., tangible objects and physical environments are acquiring a digital representation); nevertheless, people's interaction with their environment will not cease to be goal-oriented and task-centric. However, we expect that ubiquitous computing technology will allow people to carry out new tasks, as well as old tasks in new and better ways. People will realize their tasks using the services offered by ambient ecologies. Knowledge will exist both in people's heads (in the form of upgraded skills), the ambient ecology and the AmI environment (in the knowledge bases of the artifacts). In most cases, successful realization of tasks will require the knowledge-based adaptation of task models in the changing context, because it depends on the quality of interactions among artifacts and among people and artifacts, as well as on the efficiency of adaptation mechanisms.

Adaptation is a relationship between a system and its environment where change is provoked to facilitate the survival of the system in the environment. Biological systems exhibit different types of adaptation. They have inspired the development of adaptive software systems, which use a mechanism similar to biological ontogenetic adaptation so as to regulate themselves and change their structure as they interact with the environment. This mechanism is based on the replacement of one component by another component, where both components share a common interface. This approach is common to autonomic systems.

In this chapter, we use the notion of activity spheres to describe the specific ambient ecology resources, data and knowledge required to support a user in realizing a specific goal. Activity spheres are discussed in Section 2. The formation of a system that realizes such activity spheres is supported by a service-oriented architecture, which is presented in Section 3. In order to achieve task-based collaboration among the heterogeneous members of an ambient ecology, first one has to deal with this heterogeneity, while at the same time achieving independence between a task description and its respective realization within a specific AmI environment. To this end, we employ ontology alignment mechanisms described in Section 4.

Our system supports adaptation at different levels. At the ambient ecology level, the system supports the realization of the same activity sphere in different AmI environments. At the same time, it will adapt to changes in the configuration of the ecology (i.e., a new device joining, a device going out of service etc). At the task level, the system realizes the tasks that lead to the achievement of user goals using the resources of the activity sphere. Another dimension of adaptation is the interaction between the system and the user. An intelligent (speech) dialogue is able to provide easily understood metaphors for allowing people to tailor and configure ambient ecologies in a semi-tacit way to their needs and mechanisms that allow the man–machine interaction to adapt to the user context and behavior. In order to achieve this, we use a set of intelligent agents to support adaptive planning, task realization, and enhanced human–machine interaction. These are discussed in Sections 5 and 6. In the following subsection, we shall firstly present a scenario illustrating the concepts and mechanisms discussed in the remainder of the chapter.

1.1 Life in Intelligent Adaptive Homes

The scenario that follows is based on the imaginary life of a user (Suki) who just moved to a home that is characterized by being intelligent and adaptive. The scenario will help to illustrate the adaptation concepts presented in the chapter. Figure 1 illustrates the AmI environment described in the scenario.

Suki has been living in this new adaptive home for the past 10 months. Suki's living room has embedded in the walls and ceiling a number of sensors reading inside temperature and brightness; more sensors of these types are embedded in the outside wall of the house. A touch screen mounted near the room entrance together with a microphone and speaker is used as the main control point. Suki can use multiple modalities in order to interact with his smart home. The most powerful ones are the touch screen and the speech dialogue system (SDS): The touch screen can display the situation of the house, the settings, and the commands Suki has given, as well as the rules inferred by the various agents. With the help of the SDS Suki can, for example, voice control all devices and services registered to the sphere and the sphere itself can pro-actively ask Suki for information needed, e.g., to make

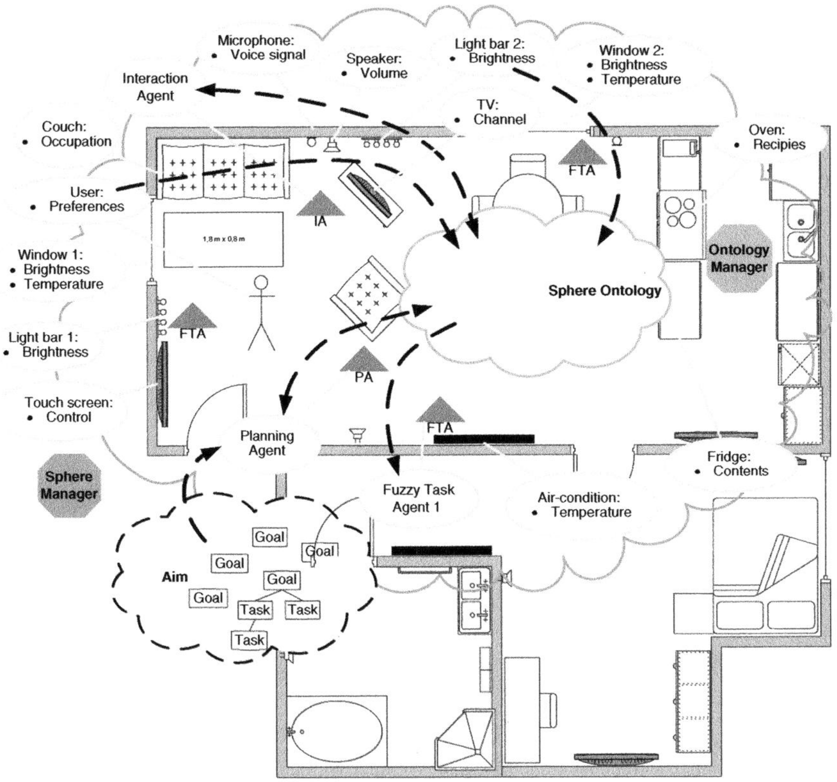

Figure 1. Elements of an activity sphere.

a system-relevant decision. The usage of these devices and services will be described in Section 6.

Suki uses an air-conditioning as the main heating/cooling device. The windows are equipped with automated blinds, which can be turned in order to dim or brighten the room. Also, the windows can open or close in order to adjust the room temperature. For the same purpose Suki can use two sets of lights in the living room. Finally, Suki has two TV sets in the house, one in the living room and one in the kitchen. The latter also contains a smart fridge, which can keep track of its contents, and an oven, which also stores an inventory of recipes and can display them in the fridge screen or the TV set. Each of these devices of the ambient ecology contains its own local ontology, which describes the device physical properties and digital services. For example, the lamp ontology stores the brand, the material, the size, as well as the location, the state (on/off) and the luminosity level. Similarly, the TV set ontology stores the set and screen dimensions, location, state, available TV channels, currently playing TV channel, statistics about channel usage, as well as viewing parameters (brightness, volume, contrast, etc.). The local ontologies will be used to form the sphere ontology, as will be described in Section 4. Moreover, these services can be directly manipulated by the task agents, as will be described in Section 5.

Suki's goal is to feel comfortable in his living room, no matter what the season or the outside weather conditions are. After careful thinking, he concluded that for him comfort involved the adjustment of temperature and brightness, the selection of his favorite TV channel, and the adjustment of volume level, depending on the TV programme.

Regarding the latter, the smart home system had observed Suki's choices over the past months and has drawn the conclusion that he tends to increase the volume when music or English-speaking movies are shown, except when it is late at night; he keeps the volume low when movies have subtitles or when guests are around. This has been possible with the help of the task agent. Nevertheless, the system does not have enough data to deduce Suki's favorite lighting and temperature conditions as the seasons change. Initially, the system will combine information in Suki's personal profile, the environmental conditions, the weather forecast, and anything else that may matter, in order to tacitly adapt to the values that Suki might want. In case of a doubt, it will engage in dialogue with Suki about specific conditions, with the help of the interaction agent. Of course, Suki can always set directly the values he desires by manipulating the devices that affect them; the system will monitor such activity and tacitly will adjust its rules. Dialogue modalities are described in Section 6.

2. Ambient Ecologies and Activity Spheres

For the intelligent ambient adaptive systems, we will introduce the ambient ecology metaphor to conceptualize a space populated by connected devices and services that are interrelated with each other, the environment, and the people, supporting the users' everyday activities in a meaningful way (Goumopoulos and Kameas, 2008). Everyday appliances, devices, and context-aware artifacts are part of ambient ecologies. A context-aware artifact uses sensors to perceive the context of humans or other artifacts and sensibly respond to it. Adding context awareness to artifacts can increase their usability and enable new user interaction and experiences.

An ambient ecology can be composed of individual artifacts and in parallel itself can be used as a building block of larger and more complex systems. Compose-ability can give rise to new collective functionality as a result of a dynamically changing number of well-defined interactions among artifacts. Compose-ability thus helps resolving both scalability and adaptability issues of ambient ecologies.

In order to model the way everyday activities are carried out within an AmI environment populated with an ambient ecology, we introduce the notion of activity sphere (Zaharakis and Kameas, 2008). An activity sphere is intentionally created by an actor (human or agent) in order to support the realization of a specific goal. The sphere is deployed over an AmI environment and uses its resources and those of the ecology (artifacts, networks, services, etc.). The goal is described as a set of interrelated tasks; the sphere contains models of these tasks and their interaction. These models can be considered as the counterparts of programs, only that they are not explicitly programmed, but are usually learnt by the system through observation of task execution. The sphere can also form and use a model of its context of deployment (the AmI environment), in the sense that it discovers the services offered by the infrastructure and the contained objects. The sphere instantiates the task models within the specific context composed by the capabilities and services of the container AmI environment and its contained artifacts. In this way, it supports the realization of concrete tasks.

Thus, a sphere is considered as a distributed yet integrated system that is formed on demand to support people's activities. An activity sphere is realized as a composition of configurations between the artifacts and the provided services into the AmI environment. People inhabit the AmI environment and intentionally form spheres by using the artifacts and the provided services. An activity sphere continuously "observes" people interactions with artifacts in different contexts, can learn their interests and habits, and can exhibit cognitive functions, such as goal-directed behavior, adaptation, and learning.

In the example we provided above, in order to satisfy the goal "feel comfortable", an activity sphere will be set up, as described in the next section. This goal can be described, for example, with abstract tasks as follows:

1 Set a comfortable temperature (TEMP)

- Sense the indoor and outdoor temperatures;
- Adjust room heating/cooling according to the user preferences and context.

2 Set a comfortable level of lighting (LIGHT)

- Sense the indoor light levels;
- Adjust indoor light levels according to the user preferences and context.

3 Select favorite TV program (FAVTV)

- Check media options;
- Set media according to the user preference and context.

The configuration of a sphere could be realized in three ways: explicit, tacit, and semi-tacit (Seremeti and Kameas, 2008). In the former mode, people configure spheres by explicitly composing artifact affordances, based on the visualized descriptions of the artifact properties, capabilities, and services. To operate this mode, people must form explicit task models and translate them into artifact affordances; then they must somehow select or indicate the artifacts that bear these affordances. The independence between object and service is maintained, although there do not exist clear guidelines regarding the degree of visibility (of system properties and seams) that a sphere should offer to people. The tacit mode operates completely transparently to the user and is based on the system observing user's interactions with the sphere and actions within the sphere. In an ideal AmI space, people will still use the objects in their environment to carry out their tasks. Agents in the intelligent environment can monitor user actions and record, store, and process information about them. Then, they can deduce user goals or habits and pro-actively support people's activities within the sphere (i.e., by making the required services available, by optimizing use of resources, etc). The sphere can learn user preferences and adapt to them, as it can adapt to the configuration of any new AmI space that the user enters. To achieve this, the encoding of task- and context-related metadata is required, as well as of the adaptation policies, which will be used by the task realization mechanisms.

The semi-tacit mode realizes a third way of configuring the sphere by combining the explicit and the implicit way. The user interacts, for example, by the use of speech dialogues with the system and provides only basic information regarding his/her goals and objectives. The user does not have to explicitly indicate artifacts and form task models but provides general commands and tells AmI some requirements he has. We assume that the semi-tacit mode may perform better than the tacit mode because the system operation within the AmI space uses the combined outcome of observation and explicit (but more general) user input. Thus it does not operate completely transparently. We assume on the one hand this is more comfortable for the user and on the other hand the system's decisions may be made closer to the user's ideas and even with a more reasonable speed.

3. System Architecture

The system we propose operates in an AmI environment, which is populated with an ambient ecology of devices, services, and people. Our basic assumption is that these components are all autonomous, in the sense that (a) they have internal models (ontologies) of their properties, capabilities, goals, and functions and (b) these models are proprietary and "closed", that is, they are not expressed in some standardized format. Nevertheless, each component can be queried and will respond based on its ontology. Finally, the AmI environment provides a registry of these components, where, for each component, its ID and address are stored.

At the same time, people have goals, which they attempt to attain by realizing a hierarchy of interrelated tasks. These are expressed in a task model, which is used as a blueprint to realize an activity sphere. Thus, for each user goal, an activity sphere is initialized, based on its task model, which consists of all software, services, and other resources necessary to support the user in achieving the goal. A sphere consists of the following (Figure 1):

1 Devices/services/AmI space properties and resources.

2 Goal and task models.

3 Software modules: Sphere manager, ontology manager.

4 Sphere ontology.

5 User(s) and user profile(s).

6 Agents: Planning agent, fuzzy systems-based task agent, interaction agent, device, or other agents (all with their local knowledge bases or ontologies).

A brief description of the components of a sphere follows.

Sphere manager (SM): The SM forms or dissolves an activity sphere and manages its several instances on different AmI spaces. It subscribes to the AmI space registry and operates as an event service for the other system components. The SM is responsible for initializing the other system components (i.e., fuzzy systems-based task(s) agent, ontology manager, etc) and thus interacts with all system components. An important role of the SM is to oversee the fulfillment of the sphere's goal. This is done in cooperation with the ontology manager which provides reasoning services. The SM also provides context-based adaptation of the activity sphere in an AmI space by deciding, for example, the replacement of a device because of a user location change that affects the task operation. The SM could be viewed as the "operating system" of a sphere virtual machine.

Ontology manager (OM): The OM aligns and merges local device, agent, policy, and user ontologies according to the task model that realizes the sphere goal. The OM is responsible for creating, dissolving, and generally managing the sphere ontology and for responding to queries regarding the sphere ontology. To that end, the OM maintains rules and provides inference services. The OM interacts with all system components.

Fuzzy systems-based task agent (FTA): One or more FTA (depending on the goal complexity) oversee the realization of given tasks within a given AmI space. These agents are able to learn the user behavior and model it by monitoring the user actions. The agents then create fuzzy-based linguistic models which could be evolved and adapted online in a life-learning mode. The FTA maintains its own local knowledge base, which is initially formed by the SM, based on the task model and the sphere ontology. New rules generated are exported to the sphere ontology through the OM. The FTA not only interacts with the SM and the interaction agent but also with the devices and services in the AmI space.

Interaction agent (IA): The IA provides a multimodal front end to the user. Depending on the sphere ontology it optimizes task-related dialogue for the specific situation and user. The IA may be triggered by both the FTA and the planning agent to retrieve further context information needed to realize and plan tasks by interacting with the user.

Planning agent (PA): The PA ensures that all tasks in the task model are described in a concrete and realizable manner and that their realization is feasible given time and resource constraints. It interacts with the SM, the OM, and the IA to enable the sphere to deal with near real-time, user-centered planning. It provides plan repairing in case a context-based adaptation by SM cannot be performed.

Sphere ontology (SO): Contains all the knowledge and data that pertain to the sphere. It is the glue that keeps these components functioning together as long as the purpose of the sphere lasts (then it is dissolved). All component interactions take place via or with the help of the ontology. It is formed on demand first by matching and then by merging or aligning the following constituent ontologies (Figure 2):

1 User profile ontologies.

2 Policy ontologies.

3 Agent local ontologies.

4 Device/service proprietary ontologies.

These ontologies are more analytically described in the following:

Device/service ontologies: They are local to each device/service. They are proprietary. We assume they do not follow a standard structure or adhere to a general formal ontology (GFO) or upper ontology. They are maintained by device/service and can be queried in a standard way.

FTA ontology: It is isomorphic to the FTA knowledge base and is maintained by the agent. Initially, it is formed by querying the SO on the devices and services that offer the services necessary for realizing the tasks in the task model. When the knowledge base changes, the agent updates the SO.

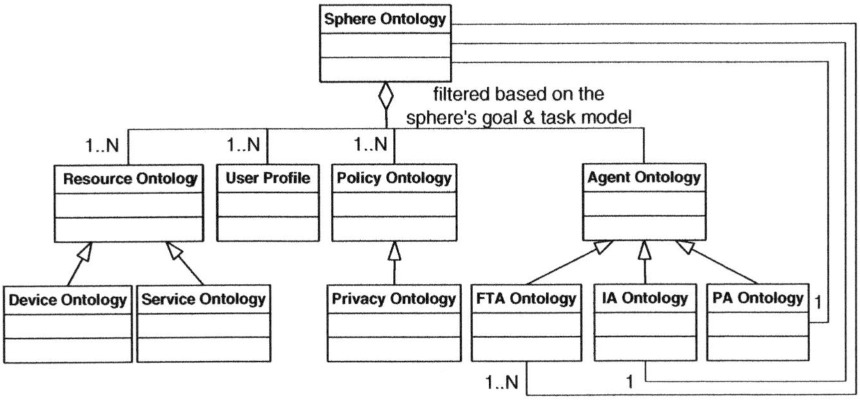

Figure 2. Domain model of sphere ontology.

IA ontology: It is maintained by the IA. Initially, it is formed by querying the SO on the devices and services that offer interaction services, based on the agent interaction policy and the tasks in the task model.

Policy ontologies: They encode entities and rules that describe specific policies, such as user privacy, multi-user conflict resolution, social intelligence, goal fulfillment conditions, and ontology dissolution. They are considered as part of the infrastructure.

User profiles: They encode user traits and preferences. A user can assume different personas based on context. A user profile can be created/updated by the FTA after monitoring device/service usage.

We assume an architecture that is service-oriented and enforces a clean service-oriented design approach, with a clear distinction between interfaces and implementation. This is similar to assumptions of component-oriented software development, as a result of which many application component platforms seamlessly incorporate service-oriented features.

When a new activity sphere is formed, the SM initializes all required components. The following steps are followed:

1 Download goal and task models from library or user profile.

2 Search for AmI space registry.

3 Instantiate PA, which tries to resolve any abstract task descriptions using the registry.

4 Initialize OM, which forms the SO based on the concrete task model and the registry.

5 Instantiate FTA and create an initial rule set using the SO.

6 Instantiate IA and create its local ontology based on interaction policy and SO.

A most common event during the operation of an activity sphere is a change in the registry of the AmI space. This may happen as a consequence of a device or service becoming unavailable, a new device arriving, or even having to instantiate the activity sphere in a new AmI space. In order to adapt to such an event, the system operates as follows:

1 The SM continuously polls for these kinds of changes and when one happens, it creates an event.

2 The PA recognizes the event and recalculates the task model.

3 The OM recognizes the event and starts a new ontology alignment process.

4 The FTA knowledge base is updated, thus the agent can now access the new configuration of the ecology.

5 The IA local ontology is updated regarding changes in the devices offering interaction capabilities.

Returning to the example, the abstract task TEMP could be made concrete by the PA with the use of devices that have been discovered in the ambient ecology. Then, the concrete task may look like the following:

1 FTA senses indoor temperature using Sensor S1 in the living room.

2 FTA senses indoor temperature using Sensor S2 in the bedroom.

3 FTA senses outdoor temperature using Sensor S3.

4 FTA checks Suki's temperature preferences stored in his local ontology.

5 FTA deduces Suki's favorite temperature for the several rooms.

6 FTA adjusts temperature by using the windows.

7 FTA adjusts temperature by using the radiator.

8 IA provides control dialogues (using different modalities) for allowing Suki to directly adjust the temperature.

Based on this description, the ontology manager will create the sphere ontology, as described in the next section, and will create the first version of the FTA knowledge base. Then, the FTA will assume direct control over the devices, monitor their usage, and update its knowledge base and its local ontology.

4. Using Ontologies to Support Adaptation

When realizing ambient spheres, one faces the challenge to develop mechanisms to support the communication between the heterogeneous devices, so as to facilitate the realization of the user's goal supported by the sphere.

In the case of activity spheres, there are multiple causes of heterogeneity:

- artifacts (devices or services) are expected to come with a proprietary, usually closed, model of itself and the world;
- intelligent environments will have their own models of their resources and services;
- user goal and task models as well as policies (i.e., for interaction, privacy, etc) will be expressed in various domain-dependent notations;
- networking and content exchange protocols usually have a restricted closed world model.

One can expect that each device in the ambient ecology will contain at least a description of its services using some popular protocol (i.e., UPnP), or even more, a set of meta-data describing its properties and services. Thus, by manipulating these local "ontologies", one can deal with the problem of heterogeneity, as well as with the problem of representing state changes.

An ontology is usually defined as "a formal, explicit specification of a shared conceptualization" (Gruber, 1993). A "conceptualization" refers to an abstract model of some phenomenon in the world, which identifies the relevant concepts of that phenomenon. "Explicit" means that the type of concepts used and the constraints on their use are explicitly defined. "Formal" refers to the fact that the ontology should be machine readable. "Shared" reflects the notion that an ontology captures consensual knowledge, that is, it is not private of some individual, but accepted by a group. Thus, an ontology is a structure of knowledge, used as a means of knowledge sharing within a community of heterogeneous entities.

Currently, there are two major standardization efforts under way in the ontology domain, carried out by IEEE and the World Wide Web Consortium. The former is concerned with a standard for upper ontology, and due to its general approach is likely to have only a limited impact. The proposal of W3C and its ontology task group resulted in the ontology language OWL (Web Ontology Language), which is the evolution of DAML+OIL. The OWL language provides support for merging of ontologies, through the use of language features which enable importing other ontologies and enable expression of conceptual equivalence and disjunction. This encourages the separate ontology development, refinement, and re-use.

The issue we face when building an activity sphere is more complex. Although common ontologies can serve as the means to achieve efficient communication between heterogeneous artifacts, it seems that they are not always effective (i.e., using a common ontology is not possible in the case where artifacts use closed proprietary ontologies). A different ontology-based mechanism is required, which will make the ontologies of the interacting artifacts semantically interoperable.

Ontology matching is the process of finding relationships or correspondences between entities of two different ontologies. Its output is a set of correspondences between two ontologies, that is, relations that hold between entities of different ontologies, according to a particular algorithm or individual. Current techniques for ontology matching require access to the internal structure of constituent ontologies, which must be verified for consistency, and result in static solutions (a set of mappings or a new ontology), which have to be stored somewhere. But an activity sphere is a transitory, dynamically evolving entity, composed of heterogeneous, independent, usually third-party components. That is why we are applying the ontology alignment technique. According to Euzenat and Schvaiko (2007), the ontology alignment process is described as follows: given two ontologies, each describing a set of discrete entities (which can be classes, properties, rules, predicates, or even formulas), find the correspondences, e.g., equivalences or subsumptions, holding between these entities. Based on these alignments, one can apply ontology merging in order to produce the top-level sphere ontology, which realizes an activity sphere.

In the example, based on the concrete task plan, which details the entities that must be used in order to realize an abstract task, the ontology manager forms the sphere ontology, which contains information about the following:

- the states of the devices and services that participate in the task;
- the knowledge bases of the sphere agents;
- the user profile;
- the constraints and policies that apply to the realization of the goal and its tasks.

5. Realizing Adaptation Over Long Time Intervals with the Help of a Fuzzy Agent

The fuzzy task agents are able to learn the user behavior and model it by monitoring the user actions. The FTA then creates fuzzy-based linguistic models which could be evolved and adapted online in a life-learning mode. This fuzzy-based system could be used to control the environment on the user behalf and to his satisfaction. The intelligent approaches used within the agents should have low computational overheads to effectively operate on the embedded hardware platforms present in the everyday environments (such as fridges, washing machines, and mobile phones) which have small memory and processor capabilities. In addition, the intelligent approaches should allow for real-time data mining of the user data and create on-the-fly updateable models of the user preferences that could be executed over the pervasive network. Moreover, there is a need to provide an adaptive lifelong learning mechanism that will allow the system to adapt to the changing environmental and user preferences over short- and long-term intervals. In all cases it is important that these intelligent approaches represent their learnt decisions and generate the system's own rules in a form that can be easily interpreted and analyzed by the end users (Hagras et al., 2007). There is a need also to provide robust mechanisms that will allow handling the various forms of uncertainties so that the system will be able to operate under the varying and unpredictable conditions associated with the dynamic environment and user preferences.

Inhabited AmI spaces face huge amount of uncertainties which can be categorized into environmental uncertainties and users' uncertainties. The environmental uncertainties can be due to the following:

- the change of environmental factors (such as the external light level, temperature, time of day) over a long period of time due to seasonal variations;
- the environmental noise that can affect the sensors measurements and the actuators outputs;
- wear and tear which can change sensor and actuator characteristics.

The user uncertainties can be classified as follows:

- intra-user uncertainties that are exhibited when a user decision for the same problem varies over time and according to the user location and activity. This variability is due to the fact that the human behavior and preferences are dynamic and they depend on the user context, mood, and activity as well as the weather conditions and time of year. For the same user, the same words can mean different things on different occasions. For instance the values associated with a term such as "warm" in reference to temperature can vary as follows: depending on the season (for example, from winter to summer), depending on the user activity within a certain room and depending on the room within the user home and many other factors;
- inter-user uncertainties which are exhibited when a group of users occupying the same space differ in their decisions in a particular situation.

This is because users have different needs and experiences based on elements such as age, sex, and profession. For instance the users might disagree on aspects such as how warm a room should be on any given day.

Thus it is crucial to employ adequate methods to handle the above uncertainties to produce models of the users' particular behaviors that are transparent and that can be adapted over long time duration and thus enabling the control of the users' environments on their behalf.

Fuzzy logic systems (FLSs) are credited with being adequate methodologies for designing robust systems that are able to deliver a satisfactory performance when contending with the uncertainty, noise, and imprecision attributed to real-world settings (Doctor et al., 2005). In addition, an FLS provides a method to construct controller algorithms in a user-friendly way closer to human thinking and perception by using linguistic labels and linguistically interpretable rules. Thus FLSs can satisfy one of the important requirements in AmI systems by generating transparent models that can be easily interpreted and analyzed by the end users. Moreover, FLSs provide flexible representations which can be easily adapted due to the ability of fuzzy rules to approximate independent local models for mapping a set of inputs to a set of outputs. As a result, FLSs have been used in AmI spaces as in Doctor et al. (2005), Rutishauser et al. (2005), Hagras et al. (2007).

Recently, type-2 FLSs, with the ability to model second-order uncertainties, have shown a good capability of managing high levels of uncertainty. Type-2 FLSs have consistently provided an enhanced performance compared to their type-1 counterparts in real-world applications (Coupland et al., 2006; Hagras et al., 2007). A type-2 fuzzy set is characterized by a fuzzy membership function, i.e., the membership value (or membership grade) for each element of this set is a fuzzy set in [0,1], unlike a type-1 fuzzy set where the membership grade is a crisp number in [0,1] (Mendel, 2001). There are two variants of type-2 fuzzy sets – interval-valued fuzzy sets (IVFS) and generalized type-2 fuzzy sets (GFS). In an IVFS the membership grades are across an interval in [0,1] and have the third dimension value equal to unity. In the case of a GFS the membership grade of the third dimension can be any function in [0,1]. Most applications to date use IVFS due to its simplicity; however, recent work has allowed GFS to be deployed efficiently.

It has been shown that IVFS-based type-2 FLSs can handle the environmental uncertainties and the uncertainties associated with a single user in a single room environment and that type-2 FLSs can outperform their type-1 counterparts (Hagras et al., 2007). However, no work has tried to approach the challenging area of developing AmI spaces that can handle the environmental uncertainties as well as the intra- and inter-user uncertainties in an environment that has multiple rooms populated by multiple users.

There are many frameworks for dealing with uncertainty in decision-making including, primarily, those based on probability and probabilistic (Bayesian) reasoning. As an aside, we emphasize that we do not claim that fuzzy-based methods are any better or any more effective than any other uncertainty handling frameworks, rather we claim that some methods are more appropriate in certain contexts. In our experience, fuzzy methods have proven to be more effective than other methods when applied in AmI spaces. This is because the fuzzy methods provide a framework using linguistic labels and linguistically interpretable rules which is very important when dealing with human users.

We shall employ the use of type-2 fuzzy logic to model the uncertainties in such AmI spaces. Consider an example of a central heating system for a living space occupied by two users. In such a situation each user's concept of cold has to be modeled throughout the year. There will be seasonal variations affecting each user's idea of what is cold, an example of intra-user variation. Each individual user will have his notion of what temperatures constitute cold, an example of inter-user uncertainty. Modeling either of these uncertainties can be accomplished using a number of existing techniques. The novel challenge with this is to model and cope with the uncertainties created by the dynamic relationship between the interactions of multiple users, each with individual preferences that change over time. For example, Figure 3(a) and (b) shows the use of type-1 fuzzy systems to depict the differences between two users (p1 and p2) concept of cold for the spring/summer and autumn/winter periods. Figure 3(c) and (d) shows how interval type-2 fuzzy sets might model each user's concept of cold throughout the year. Figure 3(e) shows how a general type-2 fuzzy set might encompass both the inter- and intra-user uncertainties about what cold is by employing the third dimension, where the different gray levels correspond to different membership levels in the third dimension.

The embedded agents learn and adapt to the user behaviors in AmI spaces using our type-2 incremental adaptive online fuzzy inference system (IAOFIS) technique. IAOFIS is an unsupervised data-driven one-pass approach for extracting type-2 fuzzy MFs and rules from data to learn an interval type-2 FLC that will model the user's behaviors.

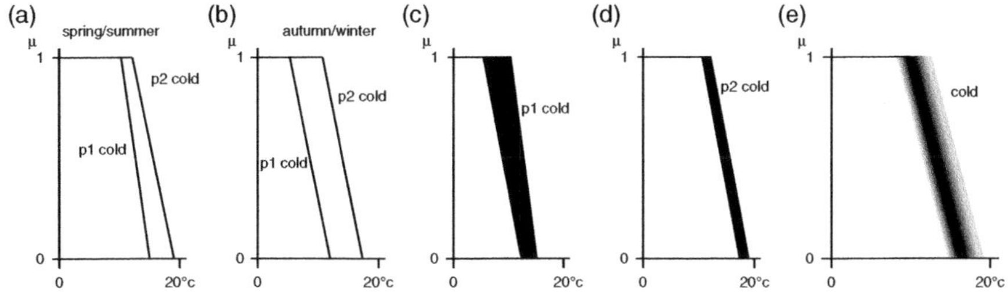

Figure 3. Fuzzy sets modeling the linguistic label cold.

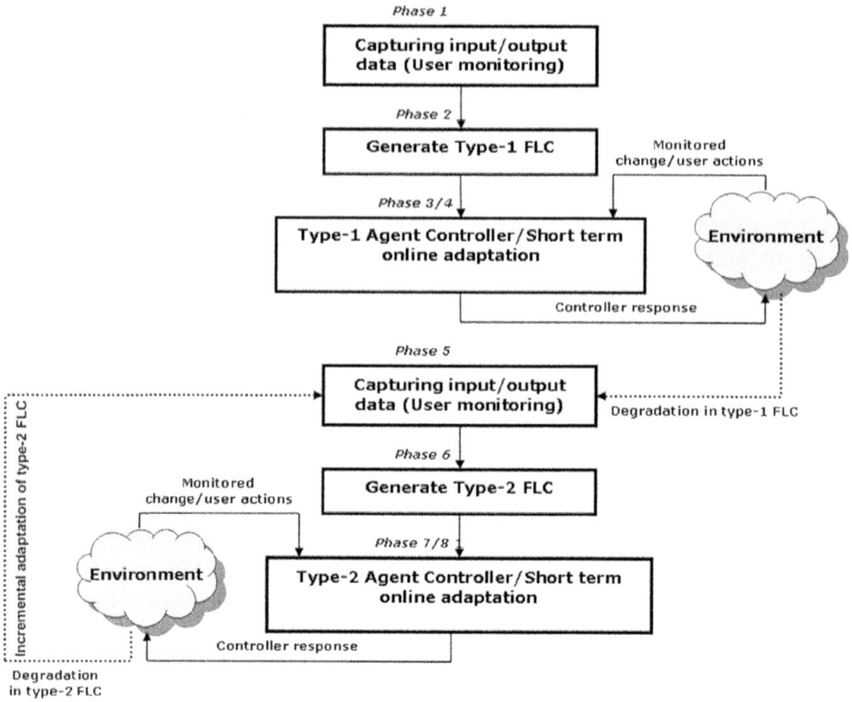

Figure 4. Flow diagram showing the main phases of IAOFIS.

The IAOFIS approach consists of eight phases of operation as described in Figure 4. In Phase 1, the system monitors the user interactions in the environment for a specific time (3 days in case of our experiments) to capture input/output data associated with the user actions. In Phase 2 the system learns from the data captured in phase 1 the type-1 MFs and rules needed to form a type-1 FLC that can effectively model the user's behaviors under the specific environmental conditions during phase 1. The approach used for learning this type-1 FLC can be found in Doctor et al. (2005), where the method for learning the type-1 MFs is based on a double clustering approach combining fuzzy-C-means (FCM) and agglomerative hierarchical clustering. In Phase 3, the learnt type-1 FLC operates in the environment to satisfy the user preferences under the faced environmental conditions. The type-1 FLC can handle the short-term uncertainties arising from slight sensor noise or imprecision as well as slight changes in environmental conditions such as small changes in temperature and light level due to variations in the weather conditions. The type-1 agent can also adapt in the short term as shown in Phase 4 by updating the FLC rule base through adding or adapting rules to reflect the user preferences associated with the encountered environment conditions. However, over a long period of time, the long-term uncertainties caused by seasonal changes and the associated changes in user activity will result in a significant deviation of the type-1 MFs parameters (associated with the linguistic labels for the input and output variables) from those initially modeled by the type-1 FLC. So whatever adaptations occur to the rules, this will not improve the system performance as the MFs values attached to the linguistic labels (which are the antecedents and the consequents of the rules) no longer reflect the current environment and user preference. This will cause the performance of the type-1 FLC to degrade which can be gauged by the increase in user interaction with the system to override the type-1 FLC outputs to try to adapt the system to his desires; this reflects the user's dissatisfaction. When the type-1 FLC sufficiently degrades a system adaptation trigger is activated and the system goes to Phase 5 in which the user is re-monitored again under the new environmental conditions for a specific time interval (again 3 days in case of our experiments). The system then goes to Phase 6 in which the agent learns the interval type-2 MFs and rules to form an interval type-2 FLC. The system then moves to Phase 7 in which the type-2 FLC controls the user environment based on the user-learnt behaviors and preferences. The type-2 agent can adapt in the short term as shown in Phase 8 by updating the type-2 FLC rule base through adding or adapting rules to reflect the user preferences associated with the encountered environment conditions. However after an extended period of time, new uncertainties arise due to the continual seasonal changes which occur in the environment, hence the type-2 MFs parameters associated with the linguistic labels change which will cause the performance of the type-2 FLC to degrade. The agent again enters a monitoring mode in Phase 5 to re-monitor the user behavior under new environmental conditions, the agent will then incrementally adapt the type-2 FLC by generating a new set of type-2 MFs and rules to take into account the current and previous uncertainties. Our system can therefore incrementally adapt the type-2 FLC in a lifelong-learning mode so that its type-2 MFs and FOUs capture all the faced uncertainties in the environment during the online operation of the agent. The adapted type-2 FLC also retains all the previously learnt user behaviors captured in the FLC rule base.

6. Adaptive User Interaction

An important dimension of adaptation is the interaction between the AmI space and the user. From a user's point of view, computer-based systems nowadays still are somehow notional and incomprehensible. In order to bridge this gap, it is necessary to establish a natural and in various ways adaptive interface between users and ambient ecologies. Displays and screens are ideally suited to give an overview about different kinds of structured information. For some other tasks we assume spoken interaction within the ambient ecology as first choice. To bring the advantages of the various ways of interaction together we consider certain rules for choosing the best input and output modality as provided:

- Output modality:
  - depends on privacy rules, e.g., somebody is entering the room;
  - depends on context, e.g., what devices are available;
  - depends on input mode;
  - depends on the information itself, e.g., music needs audio output.
- Input modality:
  - user explicitly chooses input modality;
  - depends on available resources;
  - depends on privacy rules, e.g., in public the user may not want to use speech input.

We argue that it is not a trivial problem to automatically choose the most suitable modality for interaction, whereas this choice is also an important part of adaptation. Since interaction is a huge area of research the remainder of this section exemplifies adaptive interaction by scoping on spoken interaction.

A spoken dialogue system within ambient environments may provide three classes of spoken interaction (Figure 5):

1 A main dialogue giving users the ability to control devices and services within the ambient ecology by the use of standard commands.

2 Pro-active dialogues, which are initialized by the ambient ecology.

3 On-the-fly special purpose dialogues, which are generated depending on the context and give the user a change to negotiate with the system and to ask or to submit further information.

The characteristics of the main dialogue mainly depend on the devices and services registered to the environment. Some commands to control the environment are as follows:

- control lights (e.g., "Turn on the lights!");
- control temperature (e.g., "I feel cold.");
- control blinds (e.g., "Black out the room!");
- locate things (e.g., "Where are my keys?").

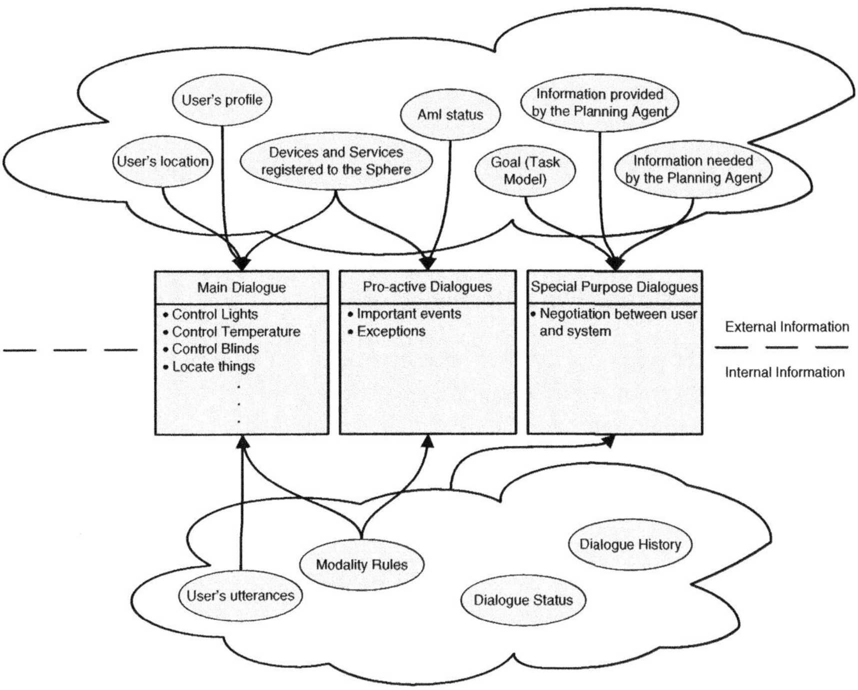

Figure 5. The main values needed for interaction and the corresponding dialogue classes.

If there are, for example, no lights available, the main dialogue will provide no commands for controlling any lighting. Thus the dialogue adapts to the infrastructure. If there are lights available an example grammar in Nuance GSL for controlling lighting may look like the following:

```
[COMMON_GRAMMAR:str]{<choice $str>}
COMMON_GRAMMAR[
[(?([turn switch] on the)lights) (lights on)] {return("lights-on")}
]
```

To increase the performance of the recognition it is possible to simplify the grammar by the use of formerly learnt "favorite commands" which may be stored in the user profile. Thus the dialogue adapts to the user. Another way of adapting may be utilized by tracking the user's location. The actual location is crucial to the main dialogue since the user's topics are in many cases directly subject to the physical surroundings. We plan to analyze if there are any other context information, e.g., time or ambience within the environment, needed for adaptive dialogue management. Apart from the mentioned external information, the main dialogue also depends on internal information such as the user's utterances and the modality rules as aforementioned.

The main dialogue could also be used for the semi-tacit configuration of the sphere mentioned in Section 2. It is imaginable to provide a main dialogue running in configuration mode by following the same ideas and using the same information resources as in standard mode.

Another class of spoken interaction are pro-active dialogues. To interact pro-actively, the ambient ecology has to decide what information is important enough to justify a pro-active system activity. The main external information underlying such a decision could be deduced from the system status and from a device or service itself. We differentiate between

- important events, e.g., AmI wakes up the user earlier because of a traffic jam to prevent him of being late at work;
- exceptions, e.g., AmI should be able to inform the user about the occurrence of exceptions which may cause damage to the environment or the residents.

The only internal information required for generating those pro-active dialogues is related to the modality rules.

Most difficult to handle is the class of special purpose dialogues. Dialogues belonging to this class give the user a change to negotiate with the system and to ask or to submit further information. The special purpose dialogues are generated depending on the external information provided by the sphere's task model itself and by the planning agent. To process a plan the planning agent on the one hand provides information which has to be communicated to the user and on the other hand requests information to accelerate planning or even to proceed with the plan. A short example scenario explains this in more detail:

Find a recipe depending on Suki's preferences and on available ingredients - Suki is hungry and wants the AmI to help him finding a recipe for a meal he would like to eat. Suki does not have an overview of the available ingredients but the Ambient Ecology does. The PA starts to find out which recipes are possible to cook with the available ingredients. It requires more information about what Suki would like and therefore a dialogue with Suki starts.

SDS: Do you rather want a warm dinner or just a sandwich?

Suki: I'm really hungry - I would like to have dinner!

To enable AmI to interact in this manner we need a more sophisticated dialogue model. In this case the internal information consists of the user's utterance, modality rules, a dialogue history, and a dialogue status. This dialogue status can, for example, follow the ideas mentioned in Larsson and Traum (2000).

To stay compatible with different well-established speech dialogue systems (TellMe, Voxeo, Nuance) we can generate dialogues using VoiceXML (Oshry et al., 2000) since this is the most common description language for speech dialogues. To minimize the limitations of VoiceXML we can use concepts similar to the popular AJAX web technologies to establish dynamic dialogues.

7. Conclusion

In this chapter we presented an architecture that can support adaptation of an intelligent environment to the requirements of specific user tasks. We model user tasks that serve a specific goal as an activity sphere and we consider that the system supporting the sphere must execute its tasks by using the services provided by the devices of an ambient ecology – we use this term to describe devices that can communicate and collaborate on demand.

The proposed architecture uses an ontology as the centralized repository of knowledge and information about the ambient ecology and a set of intelligent agents, which access and manipulate the ontology. The ontology is formed by aligning the local ontologies (or meta-data) of the devices with the concrete task descriptions, the user profile, and any policy ontologies possessed by the agents or the environment. The ontology is managed by the ontology manager and the whole activity supporting system is managed by the sphere manager.

Three kinds of adaptation can be achieved:

- task adaptation, whereby the fuzzy task agent monitors the way the user interacts with the devices of the ambient ecology and adapts its rules (and the ontology) accordingly;
- plan adaptation, in the case that the configuration of the ambient ecology changes, whereby the planning agent re-computes the concrete task descriptions based on the new configuration – plan adaptation requires the re-alignment of the sphere ontology;
- interaction adaptation, whereby the interaction agent, based on the ontology, calculates on a case basis the best way to interact with the user.

Ontology-based pervasive systems have already been presented in the literature. Among these

- the CADO (context-aware applications with distributed ontologies) framework (De Paoli and Loregian, 2006) relies on distributed ontologies that are shared and managed in a peer-to-peer fashion, so as to ensure semantic interoperability via the process of ontology merging;
- CoBrA (context broker architecture) (Chen et al., 2003) uses a collection of ontologies, called COBRA-ONT, for modeling the context in an intelligent meeting room environment. These ontologies expressed in the Web Ontology Language (OWL) define typical concepts associated

with places, agents, and events and are mapped to the emerging consensus ontologies that are relevant to the development of smart spaces;

- GAIA (Roman et al., 2002) uses ontologies as an efficient way to manage the diversity and complexity of describing resources, that is, devices and services. Therefore, these ontologies are beneficial for semantic discovery, matchmaking, interoperability between entities, and interaction between human users and computers.

All these ontology-based systems use static heavyweight domain ontologies to support ubiquitous computing applications. These ontologies are used to represent, manipulate, program, and reason with context data and they aim to solve particular ubiquitous computing problems, such as policy management, context representation, service modeling and composition, people description, and location modeling. However, in the ubiquitous computing domain, it is difficult for applications to share changing context information, as they will have to constantly adapt to the changes.

Our approach is different because it is based on (a) the existence of heterogeneous smart components (devices, services) within an ambient intelligence environment, (b) the fact that these components maintain and make available local representations of their self and state, and (c) their ability to communicate and collaborate. Thus, we propose a bottom-up scheme which maintains the independence of task description from the capabilities of the ambient ecology at hand and at the same time achieves adaptation at the collective level, without the need to manipulate local device ontologies.

References

Chen, H., Finin, T., and Joshi, A. (2003). An Ontology for Context-Aware Pervasive Computing Environments. *Knowledge Engineering, Special Issue on Ontologies for Distributed Systems*, 197–207.

Coupland, S., Gongora, M., John, R., and Wills, K. (2006). A Comparative Study of Fuzzy Logic Controllers for Autonomous Robots. In *Proceedings of the International Conference on Information Processing and Management of Uncertainty in Knowledge-Based Systems (IPMU 2006)*, pages 1332–1339, Paris, France.

De Paoli, F. and Loregian, M. (2006). Context-Aware Applications with Distributed Ontologies. In *Proceedings of the 18th International Conference on Advanced Information Systems Engineering*, 869–883.

Doctor, F., Hagras, H., and Callaghan, V. (2005). An Intelligent Fuzzy Agent Approach for Realising Ambient Intelligence in Intelligent Inhabited Environments. *IEEE Transactions on System, Man, and Cybernetics, Part A*, 35(1):55–65.

Ducatel, K., Bogdanowicz, M., Scapolo, F., Leijten, J., and Burgelman, J.-C. (2001). Scenarios for Ambient Intelligence in 2010. Technical Report, IST Advisory Group Final Report, European Commission.

Euzenat, J. and Schvaiko, P. (2007). *Ontology Matching*. New York: Springer.

Goumopoulos, C. and Kameas, A. (2008). Ambient Ecologies in Smart Homes. *The Computer Journal*, advanced access published online at doi: 10.1093/comjnl/bxn042.

Gruber, T. (1993). A Translation Approach to Portable Ontologies. *Knowledge Acquisition*, 5(2):199–220.

Hagras, H., Doctor, F., Lopez, A., and Callaghan, V. (2007). An Incremental Adaptive Life Long Learning Approach for Type-2 Fuzzy Embedded Agents in Ambient Intelligent Environments. *IEEE Transactions on Fuzzy Systems*, 15(1):41–55.

Kameas, A., Bellis, S., Mavrommati, I., Delaney, K., Colley, M., and Pounds-Cornish, A. (2003). An Architecture that Treats Everyday Objects as Communicating Tangible Components. In *Proceedings of the 1st IEEE International Conference on Pervasive Computing and Communications (PerCom 2003)*, pages 115–122. IEEE CS Press.

Kameas, A., Mavrommati, I., and Markopoulos, P. (2005). Computing in Tangible: Using Artifacts as Components of Ambient Intelligence Environments. In Riva, G., Vatalaro, F., Davide, F., and Alcañiz, M., editors, *Ambient Intelligence*, pages 121–142. IOS Press.

Larsson, S. and Traum, D. (2000). Information State and Dialogue Management in the TRINDI Dialogue Move Engine Toolkit. *Natural Language Engineering*, 6:323–340.

Mendel, J. (2001). *Uncertain Rule-Based Fuzzy Logic Systems: Introduction and New Directions*. Upper Saddle River NJ: Prentice-Hall.

Oshry, M., Auburn, R., Baggia, P., Bodell, M., Burke, D., Burnett, D., Candell, E., Carter, J., McGlashan, S., Lee, A., Porter, B., and Rehor, K. (2000). Voice Extensible Markup Language (VoiceXML) Version 2.1. *W3C – Voice Browser Working Group*.

Remagnino, P. and Foresti, G. L. (2005). Ambient Intelligence: A New Multidisciplinary Paradigm. *IEEE Transactions on Systems, Man, and Cybernetics*, 35(1):1–6.

Roman, M., Hess, C., Cerqueira, R., Ranganathan, A., Campbell, R., and Nahrstedt, K. (2002). Gaia: A Middleware Infrastructure to Enable Active Spaces. *IEEE Pervasive Computing*, 1(4):74–83.

Rutishauser, U., Joller, J., and Douglas, R. (2005). Control and Learning of Ambience by an Intelligent Building. *IEEE Transactions on Systems, Man, and Cybernetics*, 35(1):121–132.

Seremeti, L. and Kameas, A. (2008). Ontology-Based High Level Task Composition in Ubiquitous Computing Applications. In *Proceedings of The*

4th International Conference on Intelligent Environments (IE08), Seattle, USA, 1–5.

Zaharakis, I. D. and Kameas, A. (2008). Engineering Emergent Ecologies of Interacting Artifacts. In Lumsden, J., editor, *Handbook of Research on User Interface Design and Evaluation for Mobile Technology*. IGI Global, 364–384.

Chapter 4

MULTIMODAL PRESENTATION OF INFORMATION IN A MOBILE CONTEXT

Christophe Jacquet, Yolaine Bourda
SUPELEC, Gif-sur-Yvette, France
{Christophe.Jacquet, Yolaine.Bourda} @supelec.fr

Yacine Bellik
LIMSI-CNRS, Orsay, France
Yacine.Bellik@limsi.fr

Abstract This chapter deals with the design of multimodal information systems in the framework of ambient intelligence. Its agent architecture is based on KUP, an alternative to traditional software architecture models for human–computer interaction. The KUP model is accompanied by an algorithm for choosing and instantiating interaction modalities. The model and the algorithm have been implemented in a platform called PRIAM, with which we have performed experiments in pseudo-real scale.

Keywords: Ubiquitous computing; Multimodality; Mobility.

1. Introduction

Users of public places often have difficulties obtaining information that they need, especially when they are not familiar with the premises. For instance, when a passenger arrives at an airport, he does not know where his boarding gate is located. In order to provide users with potentially useful information, the staff generally places *information devices* in specific locations. These can be screens, loudspeakers, interactive information kiosks, or simply display panels. For example, monitors display information about upcoming flights at an airport and maps show the location of the shops in a shopping mall.

W. Minker et al. (eds.), *Advanced Intelligent Environments*,
DOI 10.1007/978-0-387-76485-6_4, © Springer Science+Business Media, LLC 2009

However, these information sources give non-targeted, general-purpose information suitable for anyone. As a consequence, they are generally overloaded with information items, which makes them difficult to read. Yet, a given user is generally interested in only one information item: finding it among a vast quantity of irrelevant items can be long and tedious.

Indeed, it is no use presenting information that nobody is interested in. Therefore, we propose an ubiquitous information system that is capable of providing personalized information to mobile users. The goal is not to provide *personal* information, but rather to perform a *selection* among the set of available information items, so as to present only those *relevant* to people located at proximity.

For instance, monitors placed at random at an airport could provide nearby passengers with information about their flights. Only the information items relevant to people located in front of the screens would be displayed, which would improve the screen's readability and reduce the user's cognitive load.

As we have just seen, all users are faced with difficulties when they are seeking information and have to move around in an unknown environment. However, these tasks are all the more painful for people with disabilities. Indeed, classical information devices are often not suited for handicapped people. For instance, an information screen is useless to a blind person. Similarly, a deaf person cannot hear information given by a loudspeaker.

For these reasons, we focus on *multimodal* information presentation. One given device will provide information to a user only if one of its output modalities is compatible with one of the user's input modalities. This way, the system will avoid situations in which people cannot perceive the information items.

Besides, we wish to avoid any initial specific configuration of the system. In Jacquet et al. (2006), we have proposed a framework to have display screens cooperate with each other, as soon as they are placed close to one another. In this chapter, we build on this zero-configuration system and add multimodal adaptation features.

Section 2 gives a short review of related work. Section 3 introduces a new software architecture model for ambient intelligence systems, called KUP. An agent-based embodiment of this model is introduced in Section 4. In Section 5 we propose an algorithm for choosing modalities when creating information presentations. Finally, Section 6 gives the results of experiments that have assessed the benefits of using our framework.

2. Related Work and Objectives

Computers, which were initially huge machines gathered in rooms dedicated to being *computer rooms*, made their way to the desktop in the 1980s. Then, as the use of microcomputers was becoming commonplace, people started imagining systems in which computerized information would be available everywhere, any time, and not only when one was sitting at one's desk. Hence came the notion of *ubiquitous computing* (Weiser, 1993). This notion is also referred to by the terms *pervasive computing*, *disappearing computer*, and *ambient intelligence*.

As a consequence, since the mid-1990s, several research projects have attempted to provide information to mobile users. In general, the resulting systems are built around personal digital assistants (PDAs). For instance, the Cyberguide (Long et al., 1996) pioneered the field of museum tour guides, which has seen more recent contributions (Chou et al., 2005). Some of them simply resort to displaying web pages to users depending on their location (Kindberg and Barton, 2001; Hlavacs et al., 2005).

These approaches suffer from one major drawback: they force users to carry with them a given electronic device. Even if almost everyone owns a mobile phone today, it is painful to have to stop in order to look at one's phone screen, especially when one is traveling, and thus carrying luggage. For instance, if someone is looking for their boarding desk at an airport, they would find it disturbing to stop, put down their luggage, and take out their mobile phone.

A few recent systems, such as the Hello.Wall (Streitz et al., 2003), aim at using large public surfaces to display personal information. However, to respect people's privacy (Vogel and Balakrishnan, 2004), the information items cannot be broadcast unscrambled. Thus, the Hello.Wall displays cryptic light patterns that are specific to each user. This limits the practical interest of the system, which is more an artistic object than a usable interface. Similarly, the use of the ceiling to convey information through patterns has been investigated (Tomitsch et al., 2007). The concept of *symbiotic displays* (Berger et al., 2005) enables users to use public displays as they move for various applications such as e-mail reading. However, due to the sensitive nature of this application, they are obliged to blur the most private details that the user must read on another, personal device (mobile phone or PDA). This makes the solution cumbersome because using two different devices is quite unnatural.

In contrast, we do not wish to broadcast *personal* information, but rather to *perform a selection* among the whole set of available information, which limits the scope of the privacy issues. Presentation devices will provide information relevant only to people located at proximity.

We have already proposed a model and algorithms that support the use of diverse public screens to display information to several mobile users (Jacquet et al., 2006). This is a kind of distributed display environment (DDE) (Hutchings et al., 2005). However, whereas usual DDE systems are based on static configurations of screens (see, for instance, Mansoux et al., 2005), we have introduced a model in which the assignation of information to screens changes in a purely dynamic way.

In this chapter, we take the idea further, and introduce a notion of double *opportunism* when providing and presenting information. Indeed, information items are first *opportunistically* provided by the environment, before being *opportunistically* presented onto various devices.

Besides, beyond simple content layout, we wish to be able to use several modalities. This is not dealt with by DDE studies, which focus on the physical layout of *visual* items (i.e., belonging to only *one* kind of modality). Thus, we also focus on the negotiation of multimodal content between heterogeneous users and devices. This way, we explain how a given information item can be presented on a wide range of devices with various characteristics and using various modalities. This relates to the notion of *plasticity* (Calvary et al., 2002), which studies the automatic reconfiguration of graphical user interfaces across heterogeneous devices (desktop PCs, PDAs, mobile phones, projection screens, etc.).

The methods described here are close to media allocation techniques such as those exposed in Zhou et al. (2005). This chapter describes a graph-matching approach that takes into account user-media compatibility and data-media compatibility. However, in addition to these, our solution considers *user-device* compatibility: this is the key to building an opportunistic system that can use various devices incidentally encountered as the user moves.

Note that the topic here is *not* to specify a general-purpose framework for building contextual or ambient applications. Rather, the applications that it describes may be built *on top* of such existing frameworks, for instance, those described in Dey et al. (2001) or Jacquet et al. (2005).

3. The KUP Model

In this section, we introduce a conceptual model that enables applications to provide users with personalized information in public places.

3.1 Requirements

As people rapidly move from place to place in public spaces, they will not necessarily be able to perceive a given presentation device (look at a monitor or listen to a loudspeaker) at the precise moment when a given information item is made available. As a consequence, the system must ensure that this information item is presented to them *later*, when a suitable device becomes available.

This leads us to consider two unsynchronized phases:

- In a first phase, an information item is *'conceptually'* provided to the user. This does not correspond to a physical action, but rather to an exchange of data between computer systems, corresponding, respectively, to an information source and to the user. More details are given below.

- In a second phase, this information item is *physically* presented to the user, through a suitable device and modality (text displayed on a screen, speech synthesized and emitted from a loudspeaker, etc.)

To *'conceptually'* provide information to the user, the latter must be explicitly represented by a logical entity in the system. Therefore, the KUP model introduces such an entity.

3.2 Knowledge Sources, Users, and Presentation Devices

The KUP model is a software architecture model for ambient intelligence systems. It takes three logical entities into account:

- knowledge sources, for instance, the information source about flight delays at an airport. They are denoted by K_ℓ,
- logical entities representing users, denoted by U_ℓ,
- logical entities representing presentation devices, denoted by P_ℓ.

These logical entities correspond one-to-one to physical counterparts, respectively:

- the spatial perimeter (zone) in which a certain knowledge is valid, denoted by K_φ,
- human users, denoted by U_φ,
- physical presentation devices, denoted by P_φ.

Most software architecture models for HCI (e.g., MVC (Krasner and Pope, 1988), Seeheim (Pfaff, 1985), ARCH (Bass et al., 1992), and PAC (Coutaz, 1987)) rely on logical representations for the functional core and the interface only (see Figure 1). There is no active logical representation of the user. In contrast, this entity lies at the center of the KUP model (see Figure 2):

- in the first phase, a knowledge source K_ℓ sends an information item to the logical user entity U_ℓ,
- in the second phase, the user entity U_ℓ asks a presentation entity P_ℓ to present the information item. This results in a presentation device P_φ presenting the information for the human user U_φ.

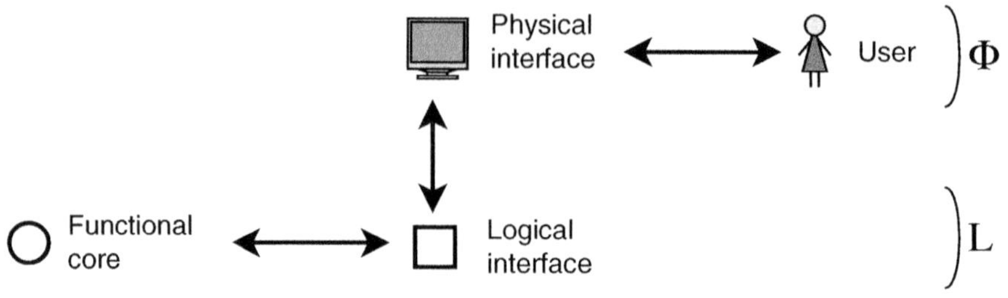

Figure 1. In classical architecture models, the user is not logically represented. The Φ and L letters, respectively, denote the physical and logical layers.

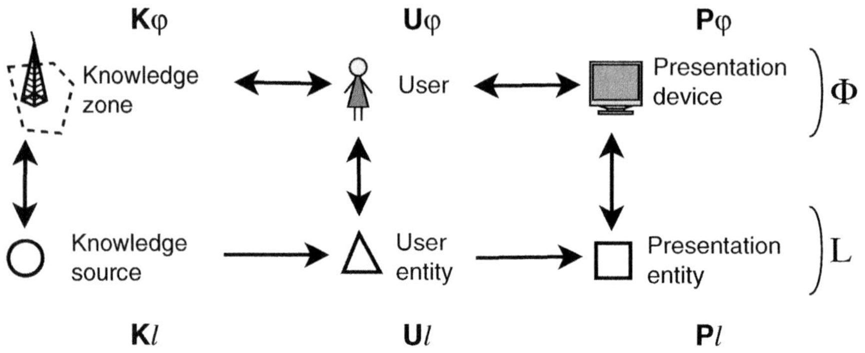

Figure 2. In KUP, a user entity lies at the center of the system. The Φ and L letters, respectively, denote the physical and logical layers.

3.3 Radiance Spaces and Perceptive Spaces

The physical entities are located in a space (denoted by $\mathcal{S}$). They have *perception* relationships with each other. Let us define these notions more formally.

3.3.1 Perceptive Space. Informally, we wish to define the *perceptive space* of a physical entity e as the set of points in space where an entity can be perceived by e. For instance, the perceptive space of a human being could coincide with his/her visual field. However, this definition is too restrictive:

1 A human being has several senses, which have various perception characteristics. For instance, the visual field of a person does not coincide with his/her auditory field. For example, a man cannot perceive a screen located 2 m behind him, but can generally perceive a sound emitted at the same place.

2 Perception depends on the *orientation* of people, which means that a point in space $\mathcal{S}$ should not only account for one's position (x, y, z) but also for one's orientation.

3 Perception depends also on the *attributes of the modalities*. For example, a phone ringing 50 m away cannot generally be heard, but a siren wailing 50 m away can be heard without any problem.

As a consequence, to precisely define the notion of perceptive space, we must take modalities and their instantiations into account. Thus, we introduce the notion of *multimodal space* or *m-space*. An m-space is the cartesian product of the space $\mathcal{S}$ with the set of all possible instantiated modalities.

For instance, let us suppose that the relevant modalities are as follows:

- a *telephone ring*, with an attribute *volume* with continuous values ranging from 0 to 100,
- a *text*, with an attribute *size* with continuous values ranging from 10 to 60 points, and an attribute *color* whose possible values are *red*, *green*, or *blue*.

Examples of points in this m-space would be

- the point at 46°23'32" N, 1°02'56" E, with a text of size 23, colored in green,
- the point at 45°07'19" N, 2°01'32" E, with a text of size 59, colored in blue,
- the point at 46°23'32" N, 1°02'56" E, with a ring of volume equal to 61.

Formally, the *perceptive space* of a physical entity e can now be defined as a subset of an m-space $\mathcal{M}$, which contains the points that e can perceive (perception being defined as above). We denote by $\mathcal{PS}(e, x)$ the perceptive space of an entity e located at $x \in \mathcal{S}$.

We have just seen that the *perception* of a physical entity e can be described in an m-space. Conversely, a physical entity can be seen as a *source of multimodal content*, whose extent is a subset of the m-space of the form $\{(x_i, m_i)\}$, where $\{x_i\}$ is the subset of $\mathcal{S}$ corresponding to the physical extent of the object, and $\{m_i\}$ the set of the entity's modalities. The set $\{(x_i, m_i)\}$ is called *location* of the entity e and is denoted by $\ell(e)$.

Note that the perceptive space depends on the particular person considered. For instance, the perceptive space of a sighted user contains the screens in front of him, located at reading distance, and the loudspeakers nearby. However, the perceptive space of a blind user located at the same place contains the loudspeakers only.

3.3.2 Radiance Space. The perceptive space of an entity describes its perception capabilities, in other terms its *behavior as a receiver.* We can now define the inverse notion, in order to describe its *behavior as an emitter* of multimodal content.

We define the *radiance space* of an entity e, with respect to an entity d, as the set of points $x \in \mathcal{S}$ from where d can perceive e, i.e., for which e is in the perceptual space of d located in x:

$$\mathcal{RS}(e|d) = \{x \in \mathcal{S} | \ell(e) \cap \mathcal{PS}(d, x) \neq \emptyset\} . \tag{4.1}$$

3.3.3 Proximity. The above definitions have introduced a notion of *proximity*. Proximity certainly depends on geographic locations, but it also depends on the multimodal capabilities of the entities. In the remainder of this chapter, we may use the terms *proximity* or *closeness* to mean *inclusion in the perceptive space*, and they must be understood as *sensory proximity.*

Proximity relationships originate in the physical world, and are then mirrored to the logical entities that are said to share the *same* relationships.

3.4 An Opportunistic and Proximity-Based Information Presentation System

Information items are formally called *semantic units.* They are elementary pieces of information, capable of being transmitted over a network, and of expressing themselves into a number of modalities.

We have seen above that there are two phases in an interaction: information *providing* and information *presentation.* The first phase can be very simple: when a user enters the perceptive space of a knowledge source, the knowledge source may send a semantic unit of interest to the logical entity U_ℓ. We will not give more details on this phase. Rather, we will focus on the second phase.

The user is mobile: when he/she receives a semantic unit, there is not necessarily a presentation device available at proximity. However, when at a given moment, one or more devices become available, the user entity will try to have the semantic unit presented on one of them. There are two interdependent sub-problems:

1 If there are several devices available, one of them must be chosen. This topic has been dealt with in Jacquet et al. (2006).

2 For a given presentation device, the user and the device must agree on a modality to be used to convey the semantic unit. Indeed, the system presented here is *multimodal* because it can successively use diverse modalities. However, it is not designed to mix several modalities to

convey one given semantic unit. This behavior is called *exclusive multimodality* (Teil and Bellik, 2000). In the future, we plan to study how to use several modalities in a complementary, redundant, or equivalent way (Coutaz et al., 1995).

The two phases that we have seen make the system's behavior opportunistic in two respects:

- with respect to information providing: the user receives semantic units when he/she enters specific areas, while moving around,
- with respect to information presentation: semantic units are presented when the user stumbles upon a presentation device.

4. Software Architecture

It would have been possible to build a system based on a *centralized* architecture. However, we think that this has a number of shortcomings, namely fragility (if the central server fails, every entity fails) and rigidity (one cannot move the knowledge sources and presentation devices at will). In contrast, we wish to be able to move, remove, and bring new entities without having to reconfigure anything. The system must adapt to the changes by itself, without needing human intervention.

That is why we propose to implement logical entities by software agents: *knowledge agents* (K), *user agents* (U), and *presentation agents* (P), respectively, associated with the logical entities K_ℓ, U_ℓ, and P_ℓ. Proximity relationships are sensed in the real world, and then mirrored to the world of agents.

We suppose that agents can communicate with each other, thanks to an ubiquitous network. This assumption has become realistic since the advent of wireless (e.g., WiFi) and mobile (e.g., GSM) networks. Besides, agents are defined as *reactive*. An agent stays in an idle state most of the time, and can react to two kinds of events:

- the receipt of an incoming network message from another agent,
- a change in its perceptive space (i.e., another agent/entity comes close or moves away).

Since all agents are only reactive, events ultimately originate in the real world. In contrast, in the real world, users are proactive[1]: they move, which is mirrored in the world of the agents, and hence trigger reactive behaviors.

The events happening in the real world are sensed by physical artifacts. For instance, RFID technology can be used to detect proximity, and hence to construct perceptive spaces. This way, monitors could detect users approaching at an airport, thanks to the passengers' tickets, provided that the tickets are equipped with RFID tags. Other possible techniques include computer vision, Bluetooth, and other wireless protocols.

5. Algorithms for Choosing and Instantiating a Modality

Our system must be capable of providing users with multimodal content. As users have different needs and wishes regarding modalities, it is necessary to choose a modality and instantiate it when interacting with a user. To begin with, we define a taxonomy of modalities.

5.1 Taxonomy of Modalities

We call *modality* a concrete form of communication using one of the five human senses (Teil and Bellik, 2000). Examples of modalities are speech, written text, or music.

Before reasoning about modality and making a choice, we have to determine the list of available modalities. Thus, we propose to build a taxonomy of modalities. Figure 3 is a partial example of such a taxonomy. It is nothing more than an example: the taxonomy can be adapted to the particular needs of any given system, enhanced, refined, etc.

In the taxonomy, all modalities are classified in a tree. Leaves represent concrete modalities, whereas internal nodes represent abstract modalities that correspond to groups of (sub-)modalities. The root of the tree is an abstract modality that encompasses every possible modality. The second-level abstract modalities correspond to human beings' senses.

This differs from Bernsen's own taxonomies of modalities (Bernsen, 1994), in which modalities are grouped according to their *arbitrary*, *linguistic*, *analogue*, or *explicit* nature, and not according to the corresponding human sense. Indeed, in our taxonomies, subsumption between modalities of the first and second levels corresponds to subsumption between sensory capabilities. However, at deeper levels, our taxonomies are closer to those of Bernsen.

Modalities have *attributes* that characterize a concrete presentation using this modality. Attributes can have discrete or continuous values. For instance, the language for a text must be selected in a finite list, whereas the text size can take any value in a given interval.

Before presenting an information item using a modality, the values for the modality's attributes have to be determined first. This step is called *instantiation* (André, 2000).

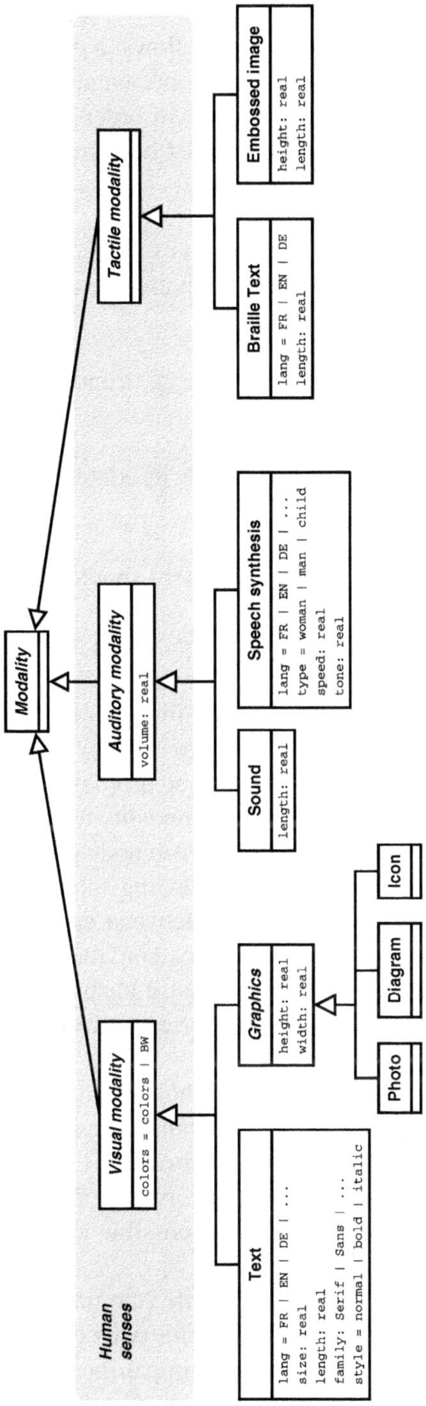

Figure 3. Example of a taxonomy of modalities.

5.2 Profiles

The problem that we have to solve is as follows: a given user wishes to have a given semantic unit presented on a given presentation device. The system must choose a modality, and instantiate it, in order to present the semantic unit. The modality and its instantiation must be compatible with each of the following:

- the user's capabilities (e.g., one cannot use a visual modality if the user is blind) and preferences (e.g., if a user prefers text to graphics, the system must try and satisfy this wish),
- the presentation device capabilities (e.g., a monochrome screen is not capable of performing color output),
- the semantic unit's capability to convey its contents using various modalities.

If there are several possibilities, the system should choose the user's *preferred solution* among them.

To solve this problem, we associate a *profile* with the user, the presentation device, and the semantic unit. These profiles describe interaction capabilities and possibly preferences, i.e., which modalities can be used, which attribute values are possible. The solution will have to comply with *each* profile, therefore it will lie at the "intersection" of the three profiles.

We define a profile as a weighting of the modality tree. A real number, comprised between 0 and 1, is associated with each node of the tree: 0 means that the corresponding modality (or the corresponding sub-tree) cannot be used; 1 means that it can be used; and values in-between can indicate a preference level. For instance, in the profile of a blind person, the sub-tree corresponding to visual modalities is weighted by 0, so that it cannot be used. Likewise, in the profile of a monitor, only the sub-tree corresponding to visual modalities is weighted by a non-null value.

The nodes' weights will determine the choice of a modality. Similarly, attributes are "weighted" too, which will help instantiating the chosen modality. More precisely, each possible value of an attribute is given a weight between 0 and 1, with the same meaning as above. Formally, a *weight function* is associated with the attribute, which maps every possible value to a weight, again a real number between 0 and 1.

Figure 4 is an example of a partial profile (the underlying taxonomy is a subset of the taxonomy of Figure 3: it contains two concrete modalities only). The profile describes a user with a visual impairment, whose native tongue is English, who speaks a little French but no German[2]. The node weights

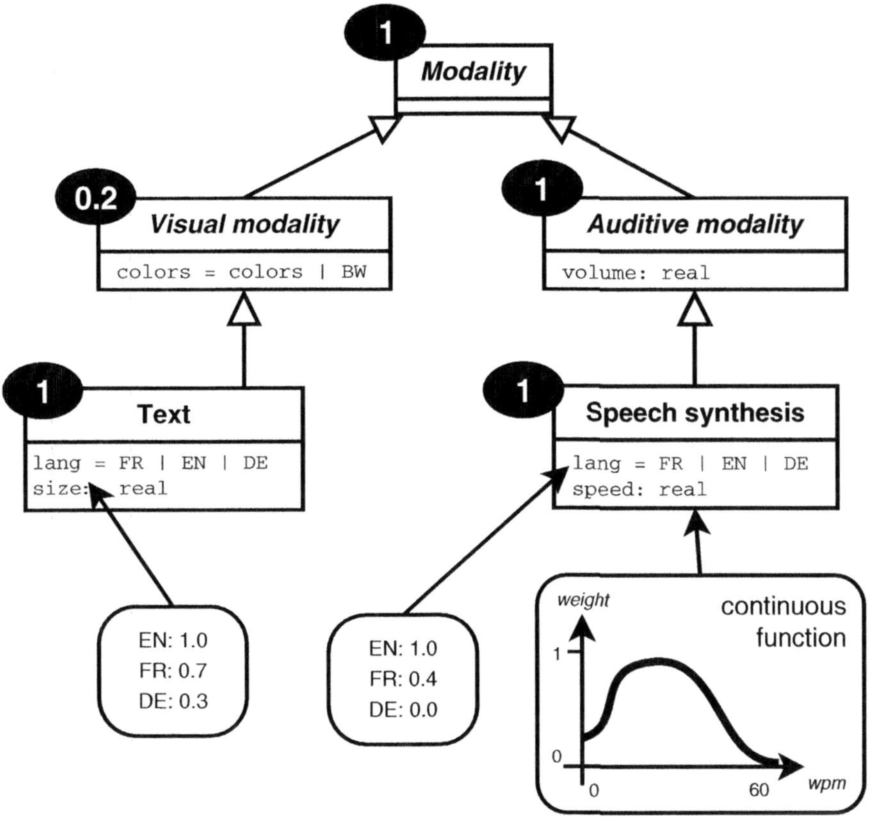

Figure 4. A partial profile (for the sake of clarity, some attribute weight functions are not shown).

are shown in white characters inside black ovals. Since the user is visually impaired, but not blind, the weight of the visual modality is low, but not zero.

The weight functions of the attributes are depicted inside boxes with rounded corners. Discrete functions are associated with attribute whose values are discrete. For instance, weights are given to any possible value of the `lang` attribute. Continuous functions are associated with attributes with continuous values. For instance, a function maps a weight to any speed, expressed in words per minute (wpm).

The examples given here are quite simple. Up to this point, we have not studied how node and attribute weights may be fine-tuned to depict real-world situations. Indeed, to take full advantage of our model, one needs methods to define weights that perfectly match the capabilities and preferences of the entities described. This issue will have to be studied from a methodological perspective.

5.3 Choosing a Modality

This section explains how the profiles can be used to determine the best possible modality instantiation when presenting semantic units. Figure 5 gives an overview of the various steps described below.

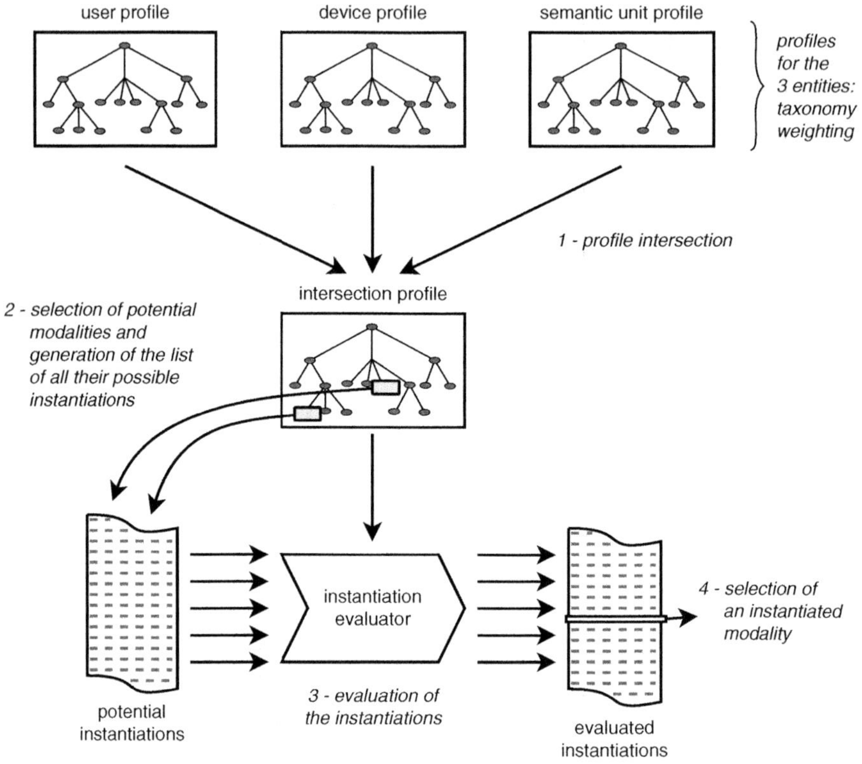

Figure 5. Overview of the algorithm for choosing a suitable modality. First, profiles are intersected, which gives out a list of usable modalities. Each possible instantiation of these modalities is *evaluated*, so as to choose the best one.

To select a modality, the system has to take the three profiles into account (user, presentation device, and semantic unit). Thus, we define the notion of the *intersection* of profiles.

The *intersection* of n profiles $p_1, \ldots, p_n$ is a profile (i.e., a weighted modality tree), in which weights are defined as follows:

- the weight of a node is the product of n weights of the same node in profiles $p_1, \ldots, p_n$,
- the weight function of an attribute is the product of n weight functions of the same attribute in profiles $p_1, \ldots, p_n$.

We call it an *intersection* because it has natural semantics. Indeed, a given node is weighted by 0 in the resulting profile if and only if there is at least one of the intersected profiles in which the given node is weighted by 0. The resulting profile is called $p_\cap$. $p_\cap$ contains information about which modalities can be used to present a given semantic unit to a given user, on a given presentation device. It also contains information to determine the values of the attributes of the chosen modality (instantiation, see below).

First, the system has to choose a concrete modality, i.e., one of the leaves of the tree. To do this, it *evaluates* each leaf. The valuation of a leaf is a real number that accounts for the weights that have been assigned to all its ancestors in the weighted tree. If an internal node has a null weight, it means that the corresponding sub-tree cannot be used, so all its leaves must be valued at zero. We could therefore define the valuation of a leaf to be equal to the product of all the ancestor node weights. However, in this case leaves with many ancestors would by nature be more likely be valued at low values than leaves with fewer ancestors.

To avoid this shortcoming, and to account for the various numbers of ancestors of the leaves, we define the *valuation* of a concrete modality (i.e., a leaf) to be the *geometric mean* of all its parent modalities' weights (including its own weight). More precisely, if $w_1, \ldots, w_m$ are the node weights along a path going from the root (weight w_1) to the concrete modality (weight w_m), then the valuation is

$$e = \sqrt[m]{w_1 \times w_2 \times \cdots \times w_m}. \tag{4.2}$$

From that, we decide to choose the concrete modality with the highest valuation.

Figure 6 illustrates profile intersection and modality evaluation on one simple example. In this case, the system would choose to use the modality that evaluates at 0.65.

5.4 Instantiating the Chosen Modality

Once a modality has been selected, the system has to determine values for its attributes. Of course, the weight functions of $p_\cap$ must be taken into account. Moreover, there must be a *global trade-off* between the needs and preferences of *all* the users located at a certain proximity, the capabilities of *all* the semantic units to be presented, and the capabilities of the presentation device.

For instance, let us suppose that two users each having one semantic unit displayed on a screen, as a line of text. Each of them would like this semantic unit to be displayed in the largest font size possible. However, the surface of the screen is limited, and so are the font sizes for each user. So the system must calculate a trade-off between the attribute values of the two semantic units.

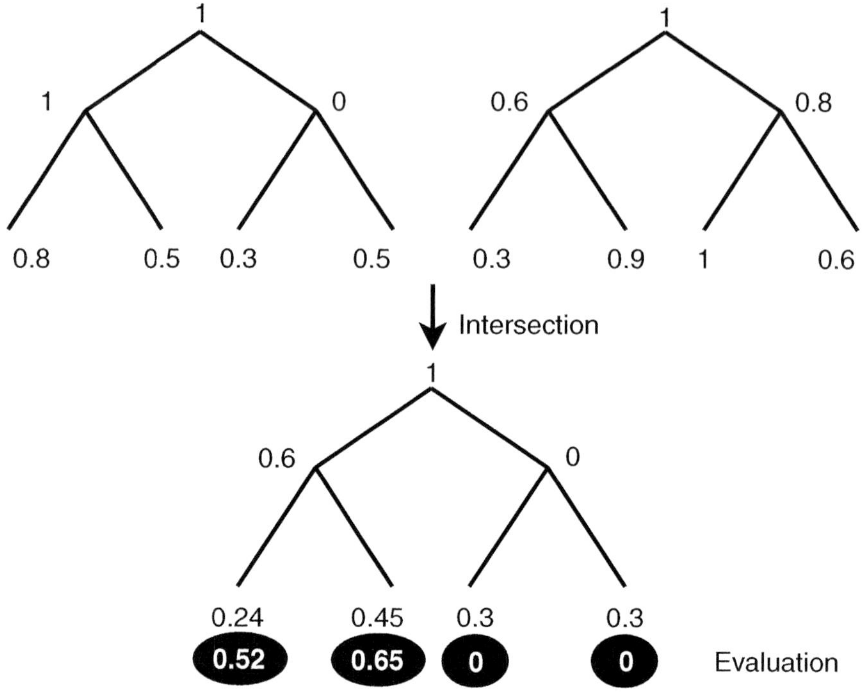

Figure 6. Intersection and evaluation.

From here on we will start reasoning at the device level. We suppose that there are a number of semantic units to be presented on a given device, which gives a total of n attributes, whose domains are called $D_1, \ldots, D_n$. We call the *attribute combination space* the set of all possible combinations of the attribute values, and we denote it by Ω. $\Omega = D_1 \times D_2 \times \cdots \times D_n$ (Cartesian product).

Some of the elements of this set are not compatible with the constraints of the presentation device. We define $\tilde{\Omega}$ as the subset of Ω whose elements are compatible with these constraints. So the "best" combination of attributes is one of the elements of $\tilde{\Omega}$. Informally, we can define the "best" solution as the solution that gives satisfaction to as many users as possible. Let us see how we can formally define this.

In a similar way as we have defined valuations above, we define the *evaluation function* of a concrete modality to be the geometric mean of the evaluation functions of the attributes of the concrete modality and its ancestors. If there are p such attributes, of domains $d_1, \ldots, d_p$ and of weight functions $f_1, \ldots, f_p$, the evaluation function of the concrete modality, denoted by e, is defined over $d_1 \times d_2 \times \cdots \times d_p$:

$$e(x_1, x_2, \ldots, x_p) = \sqrt[p]{f_1(x_1) \times f_2(x_2) \times \cdots \times f_p(x_p)}. \qquad (4.3)$$

As seen in the preceding section, for each user interested in one of the semantic units to present, there is an evaluation function. Let us suppose that there are q evaluation functions, denoted by $e_1, \ldots, e_q$. Let us take one of them, denoted by e_i. e_i is defined on a subset of $\Omega = D_1 \times \cdots \times D_n$, so it can be extended onto Ω or $\tilde{\Omega}$. We denote this extension by $\tilde{e}_i$.

Therefore, we can associate a q-component vector to each element ω of $\tilde{\Omega}$, consisting of the q values $\tilde{e}_1(\omega), \ldots, \tilde{e}_q(\omega)$ sorted by ascending order. This vector is called *valuation* of ω and is denoted by $e(\omega)$. For a given combination of attribute values, $e(\omega)$ is the list of valuations of the combination, *starting with the worst valuation.*

We want to give satisfaction to as many users as possible, so we must ensure that no one is neglected in the process. For this reason, we decide to choose the combination of attributes whose worst valuations are maximum. More precisely, we sort the vectors $e(\omega)$, for all ω, by ascending *lexicographical* order. We then choose the value ω with the greatest $e(\omega)$, with respect to this lexicographical order.

Example — let us suppose that a device has to present three semantic units for three users A, B, and C. The system has to determine the values of five attributes, given the valuations given by the three users. The results are summarized in Table 1.

Table 1. Formalization of the example situation.

ω – **Values**	e_A	e_B	e_C	$e(\omega)$ – **Valuation**
(fr, 4, de, 6, 7)	0.7	0.8	0.6	(0.6, 0.7, 0.8)
(it, 2, en, 9, 1)	0.9	0.3	0.7	(0.3, 0.7, 0.9)
(en, 2, de, 3, 5)	0.8	0.7	0.9	(0.7, 0.8, 0.9)
(es, 8, fr, 1, 3)	0.6	0.9	0.5	(0.5, 0.6, 0.9)
(de, 3, es, 7, 5)	0.2	0.4	0.95	(0.2, 0.4, 0.95)

In Table 1, the first column contains the attribute combinations. The next three columns contain the corresponding user valuations, and the last column the global valuation vector composed of the values of the three preceding columns in ascending order. The chosen solution is the third one, because it maximizes the least satisfied user's satisfaction (all user valuations are at least 0.7 in this solution).

5.5 Points of View

In the above sections, the profiles are *static*: for instance, a given user can require a minimum font size for text displays, but this size is always the same. However, depending on the distance between the user and the screen, the minimum font size should be different. Texts should be bigger, and similarly, sound should be louder as people are farther away from (respectively) a monitor or a loudspeaker.

For this reason, we introduce the notion of *points of view*. There are two possible points of view:

- *The presentation device's point of view*: Constraints on attributes (i.e., weight functions) are expressed with respect to content synthesis on the presentation devices. For instance, in this point of view, font sizes are expressed in pixels or centimeters, because these are the units used by the screen controllers to generate an image. The devices and semantic units' profiles are expressed in this point of view.
- *The user's point of view*: Constraints are expressed with respect to what is perceived by the user. For instance, from this point of view, font sizes are expressed as perceived sizes. Perceived sizes can be expressed as angular sizes (see Figure 7). In this way, measures correspond to what users can actually perceive, independently of the distance to the presentation devices. Only the users' profiles are expressed in this point of view.

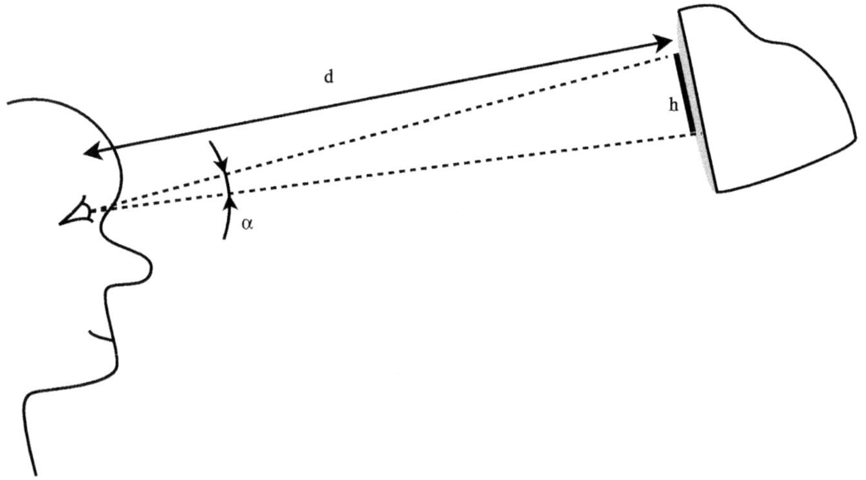

Figure 7. α is the angular size of the object of height h located on the screen at a distance d from the user.

As a consequence, the three profiles must be converted to a common point of view. As the point of view which will eventually be used to synthesize a presentation is that of the presentation device, we simply propose to convert the user profile into this one. See an example of how this works.

Let us convert a font size, expressed as an angular size in the point of view of a user, into a linear size in the point of view of a screen (see Figure 7). The user is at a distance d from the screen, the angular size is α, and the linear size is h. Then we have

$$\tan\left(\frac{\alpha}{2}\right) = \frac{h}{2d}. \tag{4.4}$$

Thus, knowing the distance d, it is very easy to translate constraints expressed in the user's point of view into the screen's point of view. Similar formulae can be found for other modalities and quantities. This allows users with sensory impairments to express constraints that will ensure that they can perceive information, regardless of their distance to the presentation devices.

6. Implementation and Evaluation

To evaluate the theories exposed above, we have implemented the models and algorithms, and then used this implementation to carry out experiments with real users.

6.1 PRIAM: A Platform for the Presentation of Information in Ambient Intelligence

We have built an implementation of the framework described in this chapter. It is called PRIAM, for PResentation of Information in AMbient intelligence. It is based on Java. Network transparency is achieved, thanks to RMI[3].

To design an application with PRIAM, one has to define classes for the agents that will provide information (K), present information (P), and model users (U). This can be done by sub-classing high-level abstract classes or simply by reusing (or adapting) classes from a library of common entities: user, information screens, simple knowledge sources, etc.

To assess the validity of our approach, we have implemented an on-screen simulator that represents an environment where people and devices are in interaction with each other (Figure 8). Proximity relationships can easily be manipulated by dragging-and-dropping objects. This has enabled us to debug and fine-tune our algorithms before conducting pseudo-real-scale evaluations.

The goal of the evaluations is to demonstrate the interest of dynamic information presentation systems for mobile users. They were conducted in our laboratory, with real users. The evaluations are based on screen displays.

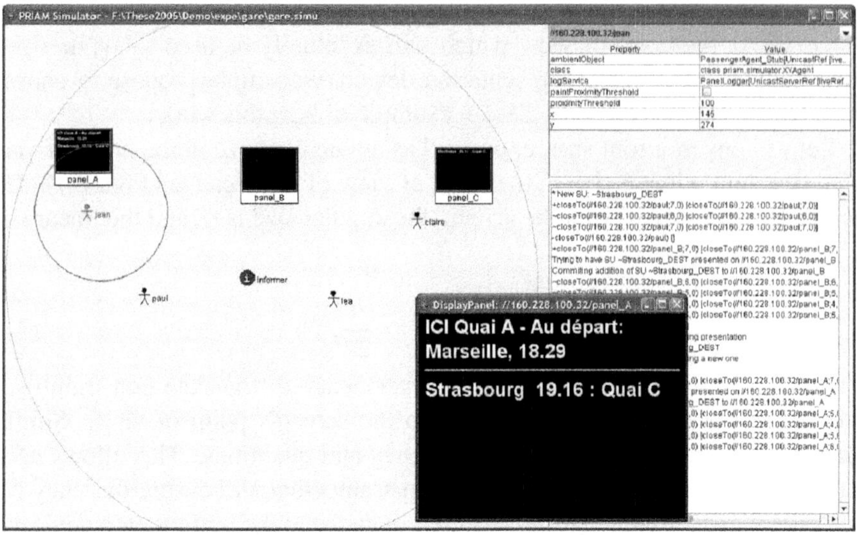

Figure 8. Screenshot of the simulator.

Proximity among screens and users can be read by sensors, thanks to infrared badges. Other techniques could have been used, such as RFID, but infrared presents a significant benefit: they not only allow the detection of people's proximity but also of people's orientation. In this way, someone who is very close to a screen, but turning his/her back to the screen, is not detected. Interestingly, this corresponds to the notion of *perceptual proximity*.

6.2 Information Lookup with Dynamic Displays

We performed an evaluation so as to assess the impact of dynamic display of information in terms of item lookup time. Sixteen subjects had to find an information item among a list of other similar items. We proposed two different tasks: to find an exam results from a list (after sitting for an exam) and to find the details about a flight. We measured the lookup time for each user, with respect to the number of users simultaneously standing in front of the list. There were one to eight simultaneous users (see Figure 9), which seems to be realistic of the maximum number of people who can gather around the same display panel.

In control experiments, users were presented with fixed-size dynamic lists, containing 450 examination marks (see Figure 10) or 20 flight details. When using the dynamic system, the display panel showed only the information relevant to people standing at proximity (i.e., one to eight items), see Figure 11.

Figure 9. Looking up exam results in a list (in the *back*) or on a screen (on the *left*). This is a picture from the experience video.

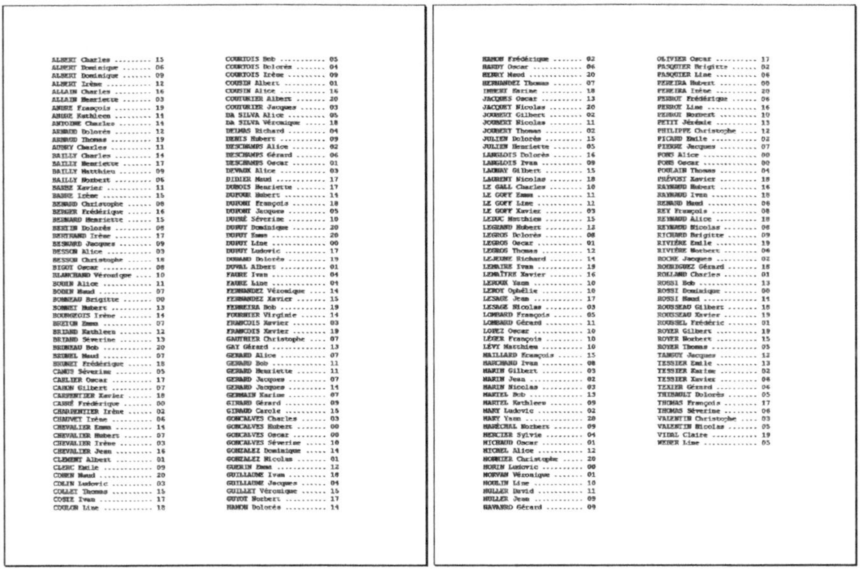

Figure 10. A static list of marks, on paper sheets.

This experiment showed that information lookup was far quicker when information display was dynamic:

- as for the exam results (see Figure 12), lookup times were 51–83% shorter (depending on the number of users), and in average 72% shorter,
- as for flight lookup (see Figure 13), lookup times were 32–75% shorter (depending on the number of users), and in average 52% shorter.

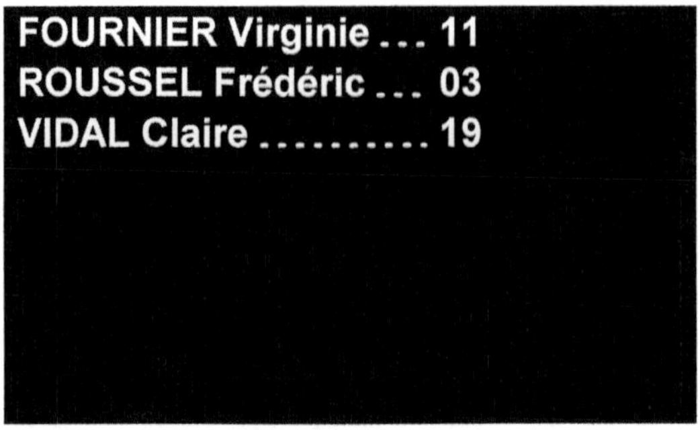

Figure 11. A dynamic list of marks, displayed on a screen.

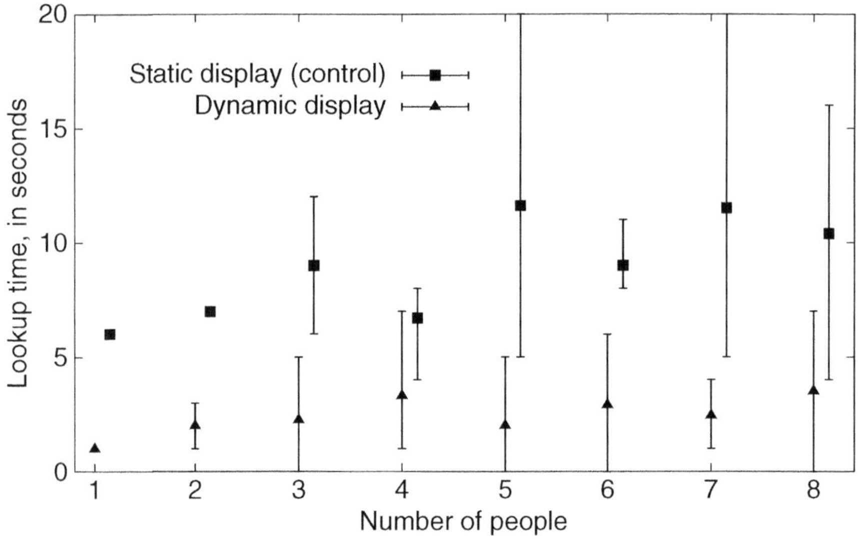

Figure 12. Mark lookup time, with respect to the number of simultaneous people. The *vertical bars* represent standard deviations, the *dots* average values.

6.3 Avoiding Unnecessary Moves in a Train Station

In a second experiment, we *added* dynamic information to initially static display screens, such as those located in train stations' subways. In a subway, a screen is located near the passageway leading to each platform: it displays the departure times for the trains on that platform. However, when a passenger changes trains, he/she initially has no clue which direction to take, so roughly half of the time, he/she first walks the whole length of the subway in the wrong direction, and then has to go back.

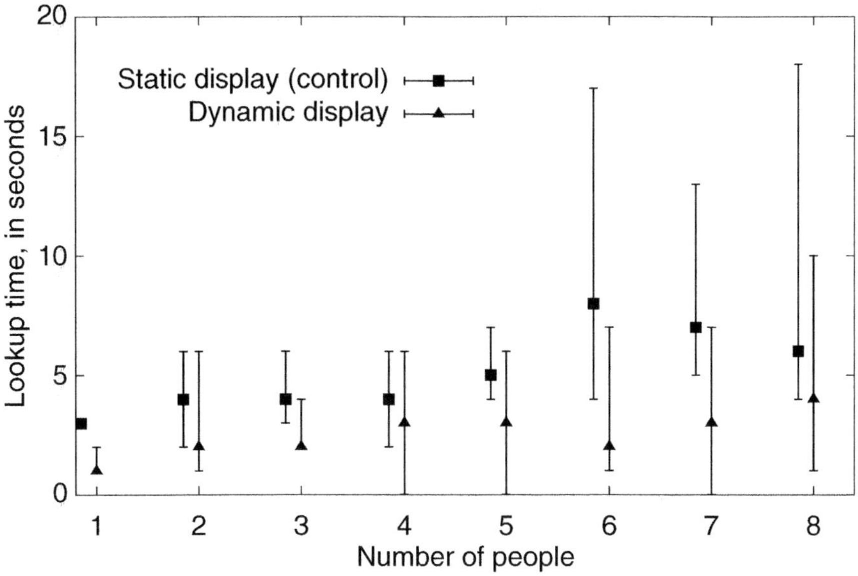

Figure 13. Flight information lookup time, with respect to the number of people present simultaneously. The *vertical bars* represent standard deviations, the *dots* represent average values.

Our idea is to display personalized information on *any* screen when a passenger approaches. This information can include the platform number as well as an arrow indicating the direction. It does not *replace* the usual static display of departing trains on the platform associated with screen, but comes *in addition* to that content. We assumed that it would help people walk directly to the right platform.

We reproduced a station subway in a corridor of our laboratory. Five display screens represented platform screens. People started from a random location in the subway, and had to take a train to a given destination, whose platform was not known by the passengers. When users had found their "platform", they had to raise their hands (Figure 14). We counted the number of elementary moves of the users (n_u), and compared it to the *optimal* number of necessary elementary moves (n_o). The ratio n_u/n_o is called the *relative length* of the paths.

When provided with static information only, people often make mistakes, which resulted in unnecessary moves (Table 2). When provided with additional dynamic information, however, they *always* followed optimal paths (relative length of 1). These results were confirmed even when several users had to go to different platforms at the same time. Moreover, people seemed to enjoy using this system, and did not feel disturbed or distracted.

Figure 14. This corridor reproduced a subway in a train station. Display screens were installed at regular intervals, along the wall.

Table 2. Relative lengths when users were provided with static information only.

Subject	n_u	n_o	Relative length
a	7	4	1.75
b	3	3	1.00
c	9	2	4.50
Average	—	—	2.42

6.4 Conclusions of the Evaluations

The evaluations have shown the benefits of dynamic display of information for mobile users. This turns out to allow very quick lookup of information on lists. Moreover, providing mobile users with supplementary personalized direction information enables a drastic decrease in the number of unnecessary moves.

However, people were generally disturbed by the items dynamically appearing and vanishing, which caused complete redisplays each time, because the lists were constantly being re-sorted. This problem could be addressed by inserting transitions when adding and removing items, or by inserting new items at the bottom of the lists instead of sorting them. Techniques for automated layout (Lok et al., 2004) and dynamic layout reorganization (Bell and Feiner, 2000) could be investigated.

7. Conclusion and Perspectives

We have presented a model and an algorithm that enable the design of multimodal information presentation systems. These systems can be used to provide information to mobile users. They intelligently make use of public presentation devices to propose personalized information. We have performed evaluations in pseudo-real conditions, which leads us to consider the following perspectives.

On a given screen, it could be interesting to *sort* the various displayed semantic units according to various criteria rather than just alphabetically or in a chronological way. A *level of priority* could thus be associated with each semantic unit. This would allow higher priority semantic units (e.g., flights which are about to depart shortly or information about lost children) to appear first. Similarly, there could be priorities among users (e.g., handicapped people or premium subscribers would be groups of higher priority). Therefore, semantic units priority levels would be altered by users' own priorities.

As seen above, priorities will determine the layout of items on a presentation device. Moreover, when there are too many semantic units so that they cannot all be presented, priorities could help choose which ones should be presented.

If a user is alone in front of a screen, then only his/her own information item is displayed, for instance, the destination of his/her plane. This can raise privacy concerns if someone is watching from behind. These issues will be the object of future work. A simple work-around would be to display one or two randomly chosen irrelevant items on the screen when only one person is present, thus confusing the malevolent persons. Or instead of displaying relevant items only, we could display all the items and then guide people's gaze to statically displayed items, thanks to personal audio clues, in a way similar to the eye-guide system (Eaddy et al., 2004).

The physical layout of semantic units (i.e., computing the positions of visual units on a screen, scheduling the temporal succession of audio units, etc.) needs to be addressed and ergonomic considerations need to be taken into account. The algorithm presented in Zhou et al. (2005) features metrics to coordinate presentations, and is therefore able to enforce ergonomic rules such as ensuring presentation ordering and maintaining presentation consistency. Implementing metrics like these would surely benefit to our system.

Our first experiments took place in *simulated* environments (a room and a corridor in our laboratory). So in the short term, we plan to carry out real-scale experiments, for instance, at an airport or train station.

Their goal will not be to test and validate the algorithms, because we have already verified their behavior with the simulator and the experiments, but rather

- to evaluate the overall usability of the system: How do users react to such a highly dynamic system? As we have seen in the experiments performed so far, some people are disturbed by too much dynamicity.
- to study the sociological impact of this system. Does it help people feel at ease when moving around unknown places, or conversely does it infringe on their privacy?
- to test the platform's usability from the point of view of the application designer: Is it easy to create an application? What are the guidelines to follow?
- in particular, the problem of assigning the weights for the modalities in the taxonomy needs to be addressed. A multidisciplinary study needs to be performed in order to assess the degree of appropriateness of particular modalities in various contexts.

Notes

1. Presentation devices and knowledge sources may be proactive too. They can be moved, yet at a different pace and rate. For instance, staff can move monitors at an airport or can change the radiance space of a knowledge source so as to reflect a new organization of the airport.
2. Only these three languages are given weights here because these are the only possible values for the `lang` attribute *in this taxonomy*. Of course, the system can easily support more languages, provided that they are defined in the taxonomy.
3. RMI: remote method invocation.

References

André, E. (2000). The Generation of Multimedia Presentations. In Dale, R., Moisl, H., and Somers, H., editors, *A Handbook of Natural Language Processing*, pages 305–327. Marcel Dekker, Inc. New York.

Bass, L., Faneuf, R., Little, R., Mayer, N., Pellegrino, B., Reed, S., Seacord, R., Sheppard, S., and Szczur, M. R. (1992). A Metamodel for the Runtime Architecture of an Interactive System. *SIGCHI Bulletin*, 24(1):32–37.

Bell, B. and Feiner, S. (2000). Dynamic Space Management for User Interfaces. In *Proceedings of the 13th Annual ACM Symposium on User Interface Software and Technology*, pages 239–248. ACM Press New York, USA.

Berger, S., Kjeldsen, R., Narayanaswami, C., Pinhanez, C., Podlaseck, M., and Raghunath, M. (2005). Using Symbiotic Displays to View Sensitive Information in Public. In *The 3rd IEEE International Conference on Pervasive Computing and Communications (PerCom 2005)*, pages 139–148.

Bernsen, N. (1994). Foundations of Multimodal Representations: A Taxonomy of Representational Modalities. *Interacting with Computers*, 6(4):347–371.

Calvary, G., Coutaz, J., Thevenin, D., Limbourg, Q., Souchon, N., Bouillon, L., Florins, M., and Vanderdonckt, J. (2002). Plasticity of User Interfaces: A Revised Reference Framework. *Proceedings of the 1st International Workshop on Task Models and Diagrams for User Interface Design Table of Contents*, pages 127–134.

Chou, S.-C., Hsieh, W.-T., Gandon, F. L., and Sadeh, N. M. (2005). Semantic Web Technologies for Context-Aware Museum Tour Guide Applications. In *Proceedings of the 19th International Conference on Advanced Information Networking and Applications (AINA'05)*, pages 709–714. IEEE Computer Society, Washington, DC.

Coutaz, J. (1987). PAC, an Object-Oriented Model for Dialog Design. In Bullinger, H.-J. and Shackel, B., editors, *Proceedings of International Conference on Human-Computer Interaction (INTERACT'87)*, pages 431–436. North-Holland.

Coutaz, J., Nigay, L., Salber, D., Blandford, A., May, J., and Richard, Y. M. (1995). Four Easy Pieces for Assessing the Usability of Multimodal Interaction: The CARE Properties. In *Proceedings of International Conference on Human-Computer Interaction (INTERACT'95)*, pages 115–120.

Dey, A. K., Salber, D., and Abowd, G. D. (2001). A Conceptual Framework and a Toolkit for Supporting the Rapid Prototyping of Context-aware Applications. *Human Computer Interaction*, 16(2-4):97–166.

Eaddy, M., Blaskó, G., Babcock, J., and Feiner, S. (2004). My own Private Kiosk: Privacy-Preserving Public Displays. In *Proceedings of Eighth International Symposium on Wearable Computers (ISWC 2004)*.

Hlavacs, H., Gelies, F., Blossey, D., and Klein, B. (2005). A Ubiquitous and Interactive Zoo Guide System. In *Proceedings of the 1st International Conference on Intelligent Technologies for Interactive Entertainment (Intetain 2005)*, volume 3814 of *Lecture Notes in Computer Science (LNCS)*, pages 235–239. Springer.

Hutchings, D., Stasko, J., and Czerwinski, M. (2005). Distributed Display Environments. *Interactions*, 12(6):50–53.

Jacquet, C., Bellik, Y., and Bourda, Y. (2006). Dynamic Cooperative Information Display in Mobile Environments. In Gabrys, B., Howlet, R., and Jain, L., editors, *10th International Conference on Knowledge-Based & Intelligent Information & Engineering Systems (KES2006)*, volume 4252 of *Lecture Notes in Artificial Intelligence (LNAI)*, pages 154–161. Springer-Verlag, Germany.

Jacquet, C., Bourda, Y., and Bellik, Y. (2005). An Architecture for Ambient Computing. In Hagras, H. and Callaghan, V., editors, *The IEE International Workshop on Intelligent Environments (IE 2005)*, pages 47–54. The IEE.

Kindberg, T. and Barton, J. (2001). A Web-Based Nomadic Computing System. *Computer Networks*, 35(4):443–456.

Krasner, G. E. and Pope, S. T. (1988). A Cookbook for Using the Model-View Controller User Interface Paradigm in Smalltalk-80. *Journal of Object Oriented Programming*, 1(3):26–49.

Lok, S., Feiner, S., and Ngai, G. (2004). Evaluation of Visual Balance for Automated Layout. In *Proceedings of the 9th International Conference on Intelligent User Interface*, pages 101–108. ACM Press, New York, USA.

Long, S., Kooper, R., Abowd, G. D., and Atkeson, C. G. (1996). Rapid Prototyping of Mobile Context-Aware Applications: The Cyberguide Case Study. In *Mobile Computing and Networking*, pages 97–107. ACM, New York

Mansoux, B., Nigay, L., and Troccaz, J. (2005). The Mini-Screen: An Innovative Device for Computer Assisted Surgery Systems. *Studies in Health Technology and Informatics*, 111:314–320.

Pfaff, G. E., editor (1985). *User Interface Management Systems: Proceedings of the Seeheim Workshop*. Springer.

Streitz, N. A., Röcker, C., Prante, T., Stenzel, R., and van Alphen, D. (2003). Situated Interaction with Ambient Information: Facilitating Awareness and Communication in Ubiquitous Work Environments. In *Proceedings of International Conference on Human-Computer Interaction.*

Teil, D. and Bellik, Y. (2000). Multimodal Interaction Interface Using Voice and Gesture. In Taylor, M. M., Néel, F., and Bouwhuis, D. G., editors, *The Structure of Multimodal Dialogue II*, Chapter 19, pages 349–366. John Benjamins Publishing Company, Amsterdam.

Tomitsch, M., Grechenig, T., and Mayrhofer, S. (2007). Mobility and Emotional Distance: Exploring the Ceiling as an Ambient Display to Provide Remote Awareness. In *The Third IET Conference on Intelligent Environments (IE 2007)*, pages 164–167.

Vogel, D. and Balakrishnan, R. (2004). Interactive Public Ambient Displays: Transitioning from Implicit to Explicit, Public to Personal, Interaction with Multiple Users. In *Proceedings of the 17th Annual ACM Symposium on User Interface Software and Technology*, pages 137–146. ACM Press, New York, USA.

Weiser, M. (1993). Some Computer Science Issues in Ubiquitous Computing. *Communications of the ACM*, 36(7):75–84.

Zhou, M. X., Wen, Z., and Aggarwal, V. (2005). A Graph-Matching Approach to Dynamic Media Allocation in Intelligent Multimedia Interfaces. In *Proceedings of the 10th International Conference on Intelligent User Interfaces (IUI '05)*, pages 114–121. ACM.

Chapter 5

CLASSIFIER FUSION FOR EMOTION RECOGNITION FROM SPEECH

Stefan Scherer, Friedhelm Schwenker, Günther Palm
Institute of Neural Information Processing, Ulm University, Ulm, Germany
{ stefan.scherer, friedhelm.schwenker, guenther.palm } @uni-ulm.de

Abstract The intention of this work is the investigation of the performance of an automatic emotion recognizer using biologically motivated features, comprising perceived loudness features proposed by Zwicker, robust RASTA-PLP features, and novel long-term modulation spectrum-based features. Single classifiers using only one type of features and multi-classifier systems utilizing all three types are examined using two-classifier fusion techniques. For all the experiments the standard Berlin Database of Emotional Speech comprising recordings of seven different emotions is used to evaluate the performance of the proposed multi-classifier system. The performance is compared with earlier work as well as with human recognition performance. The results reveal that using simple fusion techniques could improve the performance significantly, outperforming other classifiers used in earlier work. The generalization ability of the proposed system is further investigated in a leave-out one-speaker experiment, uncovering a strong ability to recognize emotions expressed by unknown speakers. Moreover, similarities between earlier speech analysis and the automatic emotion recognition results were found.

Keywords: Modulation spectrum features; RASTA-PLP; Zwicker loudness.

1. Introduction

The trend in affective computing currently aims towards providing simpler and more natural interfaces for human–computer interaction. For an efficient and intuitive interaction, the computer should be able to adapt its interaction policies to the user's emotional condition, because in "healthy" human-to-

DOI 10.1007/978-0-387-76485-6_5, © Springer Science+Business Media, LLC 2009

human interaction also two channels are important. Over the first channel explicit information using language and speech is transmitted. The second channel carries implicit information about the speaker himself, such as emotions. Therefore, it is of great interest to decode the second channel as in addition to the first in human–computer interaction. This chapter proposes the usage of three rather uncommon features for emotion recognition from speech combined by a flexible multiple classifier setup. However, recognizing emotions accurately from a speech signal, particularly in natural conversational situations is a challenging task, and so automatic detection of emotions is a rich multidisciplinary research area, which has been investigated actively in recent times (Cowie et al., 2001; Devillers et al., 2005; Fragopanagos and Taylor, 2005; Oudeyer, 2003).

Visual cues, such as facial expressions and hand gestures, are the natural indicators of emotion in human-to-human interaction. However, visual cues require additional hardware and computational resources for processing. For example, deriving relevant features from facial expressions is computationally expensive. Alternatively, vocal cues can be used for emotion recognition in applications where speech is the primary form of interaction, such as call centers and interactive voice-based systems (Cowie et al., 2001; Fragopanagos and Taylor, 2005; Scherer et al., 2003). Indications on the user's emotional status can be inferred from speech close to real-time performance by extracting simple, but informative, sets of features. Furthermore, speech-based emotion recognition can easily be incorporated into existing speech processing applications to further improve their performance and usability. The most commonly used features are pitch, energy, and spectral coefficients (Oudeyer, 2003). In this work, three rather uncommon features are utilized and their ability to detect emotion from speech is investigated.

In automatic emotion recognition from speech the features chosen to provide representative cues for the corresponding emotion are of great importance. Nicholson et al. (2000) used pitch and linear predictive coding (LPC)-based features as inputs to artificial neural networks (ANN). After the start and end point detection of an utterance, 300 different features were extracted and used as input to an ANN, resulting in the detection of eight different emotions with an accuracy of around 50% (Nicholson et al., 2000). The disadvantage of this approach clearly relies in the huge amount of features that need to be extracted from the speech signal resulting in a system far from real-time performance. Additionally, ANNs are used for classification requiring time-consuming training.

In Dellaert et al. (1996), the simple k-nearest neighbor (KNN) algorithm was used to classify four different emotions resulting in 65% accuracy. Statistics of pitch such as contour, mean, minimum, maximum, and slope were considered as features. Again the system cannot detect emotions in real time, since the statistics need to be aggregated over time and statistics over some short frames are not representative. Moreover in Scherer et al. (2003), it is mentioned that speech is a constantly changing signal and it is not sufficient to aggregate features like pitch into mean values and other statistics over the utterances in order to extract the process of emotional expression. In addition, preliminary work has shown less success in recognizing emotions with common features (Scherer et al., 2008).

In this work, features extracted from 100 ms frames of the speech signal are considered. In the earlier work by Lee et al. (2004), similar frame-based features were used as input to hidden Markov models (HMM). Mel frequency cepstral coefficients (MFCC) were extracted. The classification task was to identify four different emotions. After training, the HMMs reached an overall accuracy of around 76% (Lee et al., 2004).

In this chapter an unsupervised data reduction step, namely k-means clustering, is applied. The goal of this data reduction step is the computation of k representative prototypes for each of the seven targeted emotions. The features considered are relative spectral transform – perceptual linear prediction (RASTA-PLP) coefficients, loudness based on the work by Zwicker, and spectral energy modulation features. For each of these independent feature streams codebooks containing representative prototypes for each emotion are calculated. For classification the simple KNN classification algorithm is used as a first step. At this point the decisions are summed up to fusion over time for one utterance and for each feature stream. Furthermore, a decision fusion is realized yielding a common label for the speech signals. Using this simple approach the system can be extended easily by adding prototypes to the codebooks. Additionally, more robust features are used for emotion recognition as described in detail in Section 3.

The chapter is organized and presented in four further sections: Section 2 gives an overview of the database used for experiments, Section 3 describes the feature extraction and the classification, Section 4 presents the experiments and results, and finally Section 5 concludes by giving a summary and an outlook for further studies.

2. Database Overview

For the experiments, a standard database namely the Berlin Database of Emotional Speech is used. The data corpus comprises around 800 emotional utterances spoken in seven different emotions: anger, boredom, disgust, fear, happiness, sadness, and neutral (Burkhardt et al., 2005). Ten professional native German-speaking actors, five male and five female, participated in the recordings. Ten sentences taken from real life without emotional bias were recorded in an anechoic chamber under supervision by linguists and psychologists. The actors were advised to read these predefined sentences in the targeted seven emotions. The recordings and several data evaluations are publicly available online at `http://pascal.kgw.tu-berlin.de/emodb/`. The data are available at a sampling rate of 16 kHz with a 16-bit resolution and monochannel.

A human perception test with 20 subjects, different from the speakers, was performed to benchmark the quality of the recorded data. This test yielded a mean accuracy of around 84% (Burkhardt et al., 2005). The confusion matrix was derived for all the emotions and is shown in Table 1.

Table 1. Confusion matrix of the human performance test (in %).

	Fear	Anger	Happiness	Disgust	Neutral	Sadness	Boredom
Fear	85.3	3.5	3.1	3.9	2.8	1.3	0.0
Anger	0.7	96.0	0.8	0.8	1.3	0.2	0.2
Happiness	2.2	4.6	83.1	2.0	6.1	1.3	0.6
Disgust	3.4	2.0	1.3	79.6	3.6	8.2	1.8
Neutral	0.2	1.6	0.7	0.4	87.2	4.1	5.6
Sadness	5.8	0.3	0.4	1.7	5.6	78.4	7.9
Boredom	0.2	0.6	0.2	0.3	11.2	2.8	84.9

The values in Table 1 are listed in percentage of the utterances in one emotion. Additionally to the human recognition performance test the participants were asked to estimate the naturalness of the presented emotion. Recordings with a lower average rate of naturalness than 60% and a lower recognition rate than 80% were removed from the database leading to different numbers of available utterances in each emotion. The exact number of utterances available in all of the seven emotions is listed in Table 2. Since anger is easy to portray for actors and to recognize the number of angry utterances is much higher than the available recordings of disgust. However, anger is of great importance for popular emotion recognition tasks such as call center applications, in which often only anger needs to be recognized.

Table 2. Number of utterances with corresponding emotion available in the Berlin Database of Emotional Speech.

Fear	Anger	Happiness	Disgust	Neutral	Sadness	Boredom
68	126	71	46	78	62	80

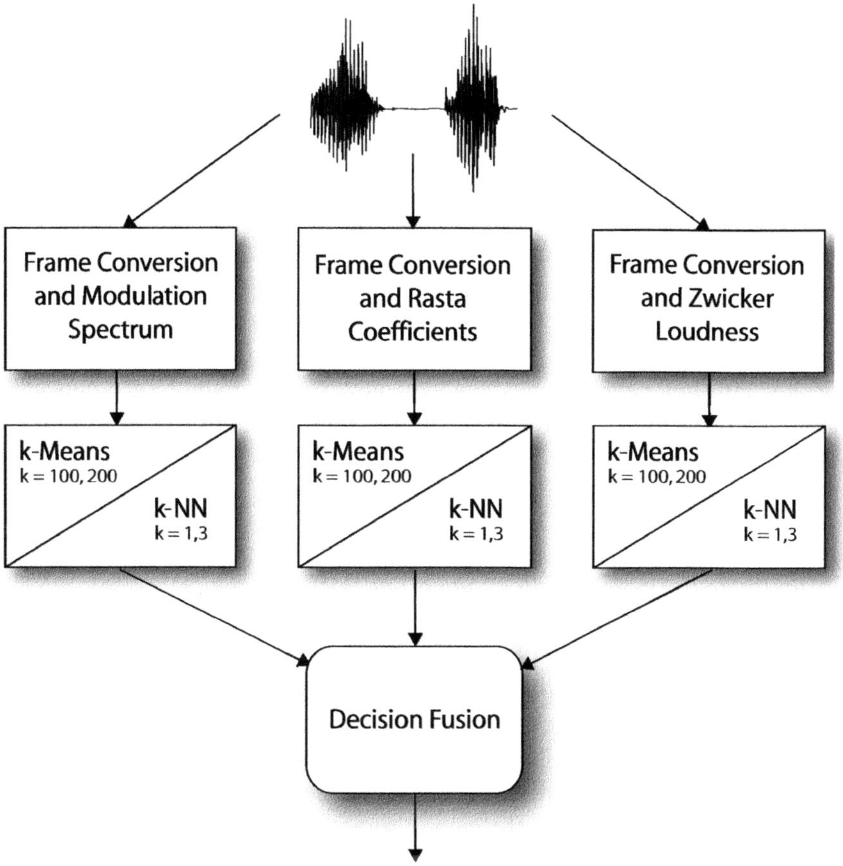

Figure 1. Algorithm overview.

3. Approach

In this work several processing steps, namely feature extraction, data reduction, classification, and decision fusion are conducted sequentially. Figure 1 shows a schematic overview, which is described in more detail in the following paragraphs. The first step, feature extraction, is needed in both training and test. The second step is divided into two, as shown in Figure 1. The upper left corner indicates the data compression step only needed during training and the lower right corner indicates the first part of the classification phase, here the KNN classification. The last step, which is called decision fusion, necessary for classification, combines the three different streams into one common output.

3.1 Feature Extraction

Feature extraction as proposed in this chapter consists of three independent feature streams, namely RASTA-PLP coefficients (Hermansky et al., 1991; Hermansky and Morgan, 1994), often used in speech recognition as an alternative to MFCC, perceived loudness as proposed by Zwicker et al. (1991), and spectral energy modulation (Hermansky, 1996). All of these features are motivated by human auditory perception, which is not equally sensitive to every frequency. Frequencies in the range of around 100–2000 Hz are perceived more intensely than higher frequencies. In the following these three features are described in more detail. Especially the modulation spectrum-based features need further explanation, since they are novel in emotion recognition, not commonly used in speech recognition, and quite difficult to understand (Krishna et al., 2007).

3.1.1 RASTA-PLP. Since conventional short-term spectrum-based speech analysis techniques do take information from irrelevant sources into account, Hermansky and Morgan developed an analysis approach suppressing spectral components that change more slowly or more rapidly than the typical range of change of speech (Hermansky and Morgan, 1994). This is possible since the vocal tract movements are reflected by the speech signal and are distinct from artificial sources influencing the signal. In their work, they have proven that RASTA processing of speech enhances the performance of automatic speech recognizers, in the presence of noise coming from high-frequency sources, such as traffic, and unforeseen changes in the recording situation, such as different microphones representing low-frequency disturbances are present. Therefore, RASTA-processed features proved to be more robust in many realistic situations (Hermansky and Morgan, 1994).

Perceptual linear prediction (PLP) is similar to the standard linear predictive coding (LPC) analysis, but respects psychophysical findings, such as the equal loudness curves mentioned in the following paragraph and critical band spectral resolution. The PLP feature extraction is based on short-term spectrum of speech and PLP is sensitive toward short-term spectral value changes. Therefore, Hermansky proposed the RASTA-PLP approach, which renders the features more robust to linear spectral distortions (Hermansky et al., 1991).

3.1.2 Zwicker Loudness. These features are based on an algorithm to extract perceived loudness (PL) proposed by Zwicker and defined in DIN 45631 (Zwicker et al., 1991). In Figure 2 the equal loudness curves are shown. One line represents the sound pressure (dB) that is required to perceive a sound of any frequency as loud as a reference sound of 1 kHz.

It is clear that emotions are mediated through loudness, e.g., anger is often expressed by using a louder voice while sadness is embodied by talking more

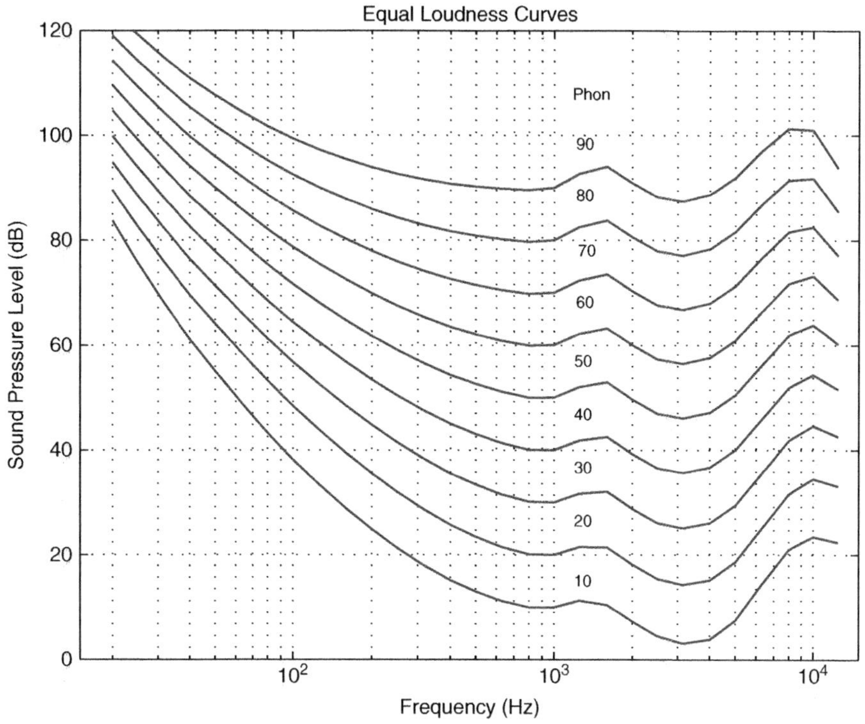

Figure 2. Equal loudness curves.

quietly or softly (Scherer et al., 2003). We have conducted a small experiment verifying if PL is capable of only separating emotions that differ from each other according to the activation dimension representing different levels of arousal in Whissel's wheel of activation-evaluation space, in which, for example, happiness, anger, fear, and disgust are classified as active, and neutral, sadness, and boredom as passive, as shown in Figure 3 (Whissel, 1989). For the given database the extracted loudness features were efficient and accurate. It was possible to separate active and passive emotions using a simple KNN classifier with an accuracy of more than 90%. Furthermore, it is possible to classify happiness vs. anger with an accuracy of around 80%, which is quite high since happiness and anger are quite difficult to separate, since they are sharing almost the same level of activation (Yacoub et al., 2003). However, in realistic applications these features may not suffice since they are strongly speaker- and microphone-dependent. One may consider a voice-based system using a global threshold for loudness indicating active emotions and a calm user calling from a noisy airport.

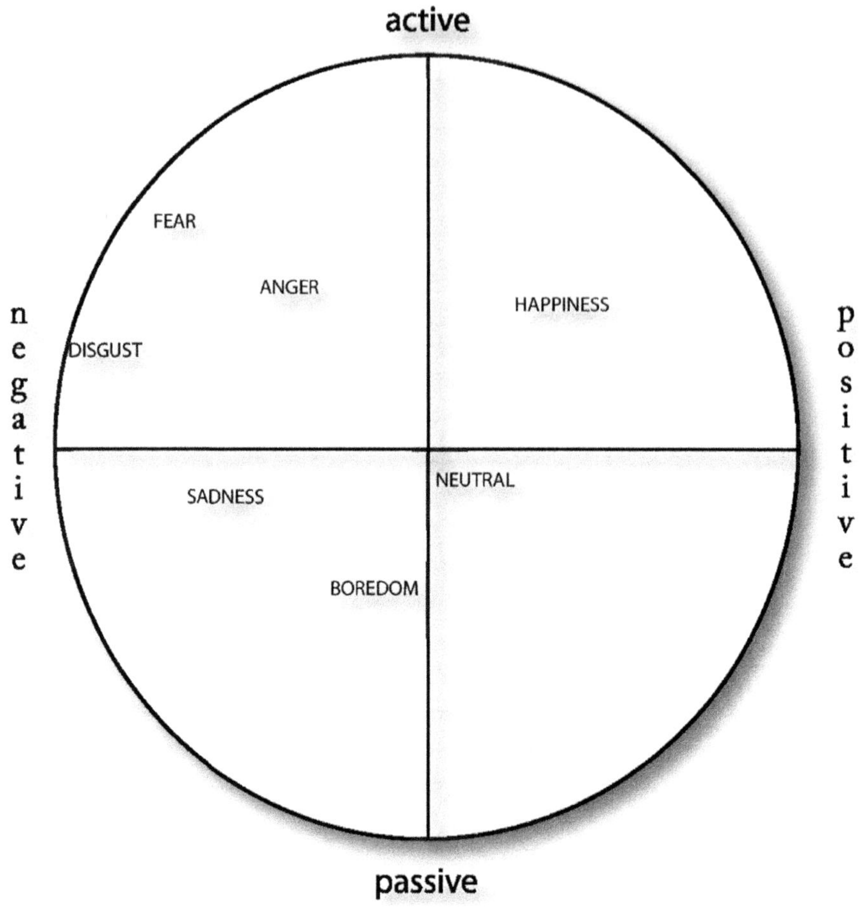

Figure 3. Activation-evaluation wheel adapted from Whissel (1989).

3.1.3 Modulation Spectrum-Based Features. The third feature type is based on long-term modulation spectrum, which describes the slow temporal evolution of speech. Again, these features emulate the perception ability of the human auditory system. Earlier studies reported that the modulation frequency components from the range between 2 and 16 Hz contain important linguistic information (Hermansky, 1996; Kanederaa et al., 1999).

This specific observation formed the basis for the proposed features to detect the emotional condition of the speaker. The brief outline of the algorithm is shown in Figure 4. In the first step, the fast-Fourier transform (FFT) of the input speech signal $x(t)$ is computed over N points in successive time windows with a shift of n samples, which results in a $N/2$ dimensional FFT vector every

Input: audio signal $x(t)$

- convert $x(t)$ into frames $x_i, ..., x_I$ of the size N with a shift of n samples
- FOR each frame x_i, $i = 1, ..., I$
  - short time fast Fourier transform X_i of x_i is computed
  - Mel-scale transformation is applied to X_i resulting in m_i^j, $j = 1, ..., 8$ bands imitating the human auditory system (see Eq. 5.3)
  - FOR each band m_i^j, $j = 1, ..., 8$
    * short time fast Fourier transform M_i^j of m_i^j over P samples is computed
    * energy $e_i^j = \sum_{p=1}^{P/2} e_i^j(p)^2$ is calculated

    END for each band m_i^j

  END for each frame x_i
- the median of each band e^j over whole utterance resulting in an eight dimensional feature vector $\hat{e}$ is calculated

Figure 4. The modulation spectrum-based feature extraction algorithm.

n time steps. Then, the Mel-scale transformation, which imitates the human auditory system, is applied to these vectors. The Mel-filter bank with $M = 8$ triangular filters $H_m[k]$ is defined by

$$H_m[k] = \begin{cases} 0 & k < f[m-1] \\ \frac{2(k-f[m-1])}{(f[m+1]-f[m-1])(f[m]-f[m-1])} & f[m-1] \leq k \leq f[m] \\ \frac{2(f[m+1]-k)}{(f[m+1]-f[m-1])(f[m+1]-f[m])} & f[m] \leq k \leq f[m+1] \\ 0 & k > f[m+1] \end{cases} \tag{5.1}$$

for $m = 1, ..., M$. The boundaries $f[m]$ are equally distributed in the Mel-scale:

$$f[m] = \frac{N}{F_s} B^{-1}\left(B(f_l) + m\frac{B(f_h) - B(f_l)}{M+1}\right), \tag{5.2}$$

with $f_l = 32$ the lowest frequency, $f_h = 8000$ the highest frequency, and F_s the sampling rate in Hz. The Mel-scale transformation is defined as

$$B(f) = 1125 \ln(1 + f/700), \tag{5.3}$$

and the inverse of the transformation as

$$B^{-1}(b) = 700(\exp(b/1125) - 1). \tag{5.4}$$

In the second step, the modulations of the signal for each band are computed by taking FFT over P points in successive time windows, shifted by p samples, resulting in a sequence of $P/2$ dimensional modulation vectors. Figure 5 shows the waveform, spectrogram, and the modulation spectrum for the utterance, "Das schwarze Blatt Papier befindet sich da oben neben dem

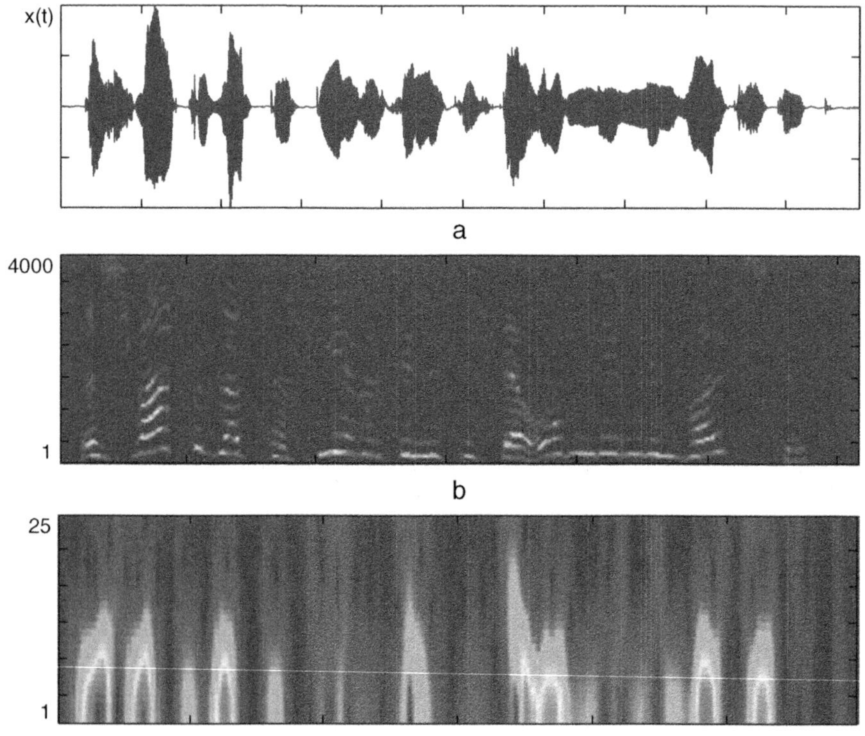

Figure 5. **(a)** Waveform **(b)** spectrogram **(c)** modulation spectrum for a given utterance. The x-axis for **(a)** is the number of samples and for **(b)** and **(c)** the number of frames. The y-axis for **(b)** and **(c)** indicate frequency in Hz.

Holzstück."[1], for the values of $N = 512$, $n = 160$, $P = 100$, $p = 1$, and $m = 5$. Such a high value for P is not suitable for the classification task, but it is necessary to exemplify how those features look like and what they represent. It is observed that most of the prominent energies are observed within the frequencies between 2 and 16 Hz as shown in Figure 5(c). Furthermore, it is seen that if there is a frequency change in the spectrogram in Figure 5(b) there is a peak in the modulation spectrogram. For the classification task following values were used: $N = 1600, n = 640, P = 10, p = 1$.

These modulation spectrum-based features were first proven efficient for emotion recognition in voice-based applications in Krishna et al. (2007). In this work a small number of features was extracted as mentioned above and used as input to a simple KNN classifier outperforming large feature sets used in earlier work (Petrushin, 1999). The task was to recognize agitated emotions, comprising anger, happiness, fear, and disgust, and calm emotions, comprising neutral, sadness, and boredom. The assignment of the seven emotions found in the Berlin Database of Emotional speech to the two categories was again according to Whissel's wheel of activation-evaluation space (Whissel, 1989). The simple architecture outperformed earlier work by more than 11% resulting in more than 88% accuracy (Krishna et al., 2007). The experiments in the present work extend the work in Krishna et al. (2007) in such a way that more emotions are classified and the combination with different features in a multi-classifier system are used.

Another indication that modulation spectrum features are suitable for emotion recognition is seen in Figure 6. The scatter plot compares the performances of modulation spectrum-based energies and pitch for anger vs. neutral classification. Pitch is extracted from the speech signal using the simple inverse filtering tracking algorithm (SIFT) proposed by Rabiner leading to integer values (Rabiner and Schafer, 1978). It is observed that modulation spectrum-based energies classify anger and neutral emotions accurately. However, with pitch, which is an important and established feature for emotion recognition, there is a huge overlap between anger and neutral emotions.

3.2 Data Reduction

Since the data output of the feature extraction is quite large and computationally expensive to deal with, a data reduction step is needed. For this purpose the well-known batch version of the k-means algorithm is utilized (Kohonen, 2001). The exact algorithm for the data reduction is shown in Figure 7. In this manner, k representative prototypes for each emotion are computed and stored in a codebook. These prototypes allow a fast, but still accurate classification.

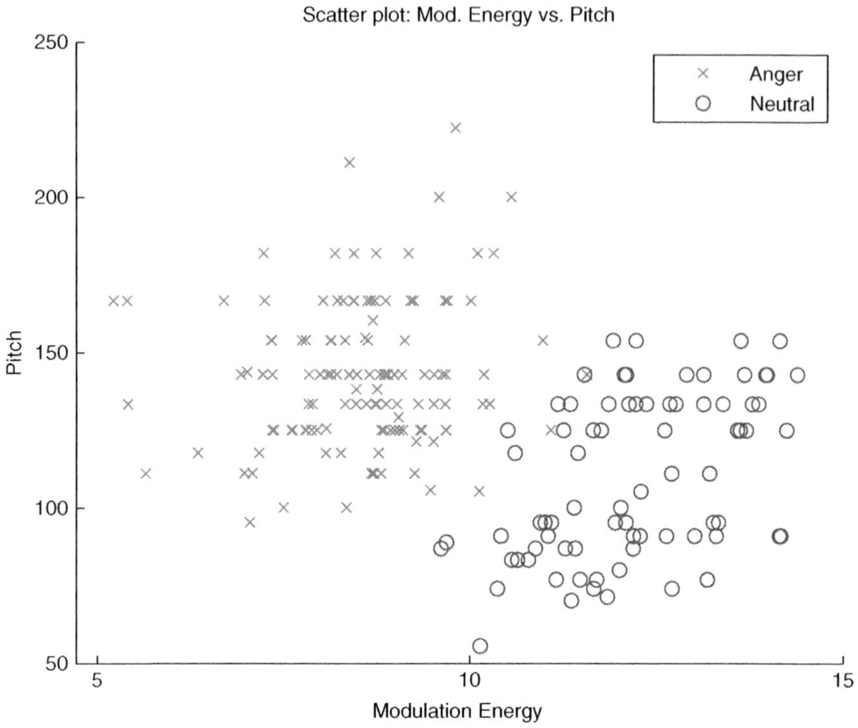

Figure 6. Scatter plot of modulation spectrum vs. pitch.

Input: feature vectors $X = \{x_1, ..., x_n\}$, set k

1 Initialize partition $C = \{C_1, ..., C_k\}$ of X randomly

2 Calculate centroids $c_i(t) = \frac{1}{|C_i|} \sum_{x_j \in C_i} x_j$ of the clusters $C_1, ..., C_k$

3 Identify new partition C by finding the closest cluster center c_i for each x_j

4 If $c_i(t) = c_i(t-1) \; \forall \, i$, then output $c_1, ..., c_k$, else go to (2)

Figure 7. Unsupervised batch k-means algorithm.

3.3 Multi-Classifier Systems

Multiple classifier systems (MCS) are special types of classifiers that integrate several classifiers for the same pattern recognition problem. The main goal of MCS is to obtain a combined classifier computing a more accurate and robust classification. In a typical scenario of MCS a complex high-dimensional pattern recognition problem is decomposed into smaller subproblems for which solutions can be simpler achieved.

In Figure 8 a diagram of the most popular type of MCS is shown. Here it is assumed that the raw data X originates from an underlying source, but each classifier receives different subsets of X, e.g., X is applied to multiple types of feature extractors $F_1, \ldots, F_N$ computing multiple views $F_1(X), \ldots, F_N(X)$ of the same raw input data X. Feature vectors $F_j(X)$ are used as the input to

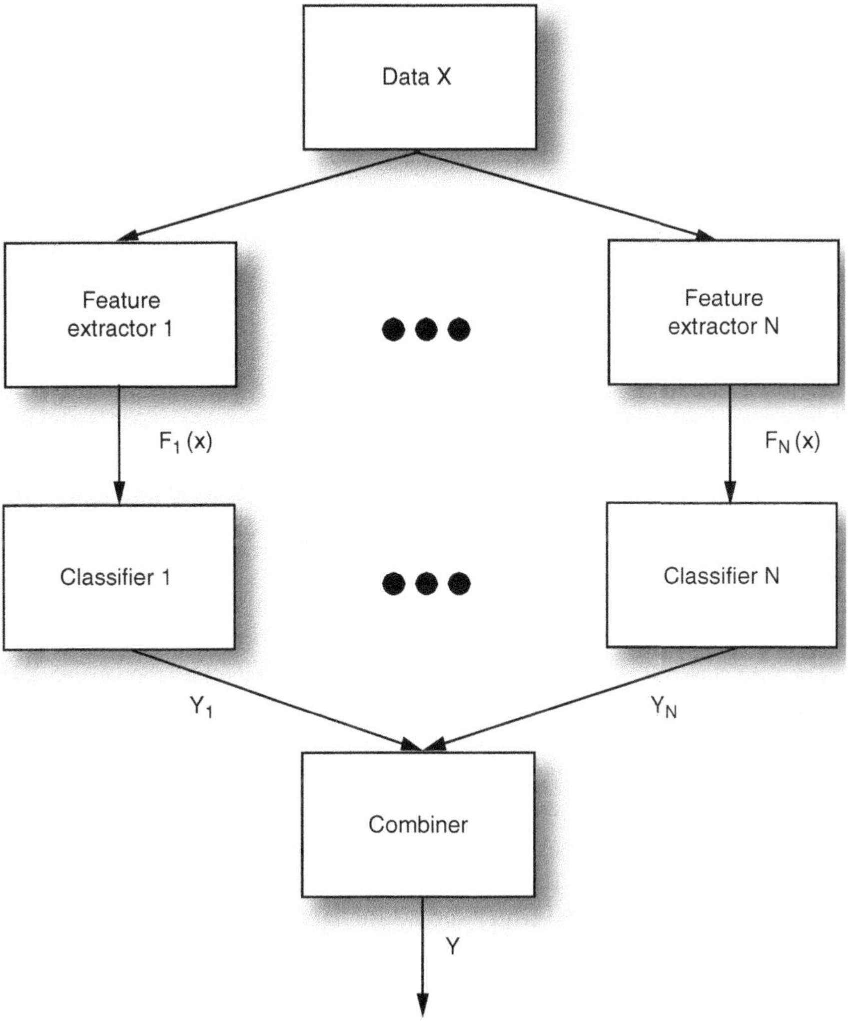

Figure 8. A generic architecture of a multi-classifier system consisting of M classifiers and feature extractors. Classifier outputs are fused into a final decision through a combination mapping.

the jth classifier computing an estimate y_j of the class membership of $F_j(X)$. This output y_j might be a crisp class label or a vector of class memberships, e.g., estimates of a posteriori probabilities. Based on the multiple classifier outputs $y_1, \ldots, y_N$ the combiner produces the combined certainty values and the final decision y. Combiners used in this study are fixed transformations of the multiple classifier outputs $y_1, \ldots, y_N$. Examples of such combining rules are voting, (weighted) averaging, and multiplying, just to mention some of the most popular types. In addition to a priori fixed combination rules the combiner can be a parameterized mapping trainable through an additional optimization procedure where the classifier outputs are then used as inputs, and the original class label as the target output of the combiner. This type of combiner may use artificial neural networks, decision templates, or support vector machines (Kuncheva, 2004).

In this work three classifiers with different input data are combined to one MCS as shown in Figures 1 and 8. Simple classifiers may not be capable of classifying several utterances correctly, but a combination of two or more leads to better classification results. Therefore, an MCS clearly aims toward more accurate classification results at "the expense of increased complexity" (Kuncheva, 2004). However, the usefulness of those more complex classifiers is not fully agreed upon. For instance, Ho critically states that by combining classifiers and combination methods over and over again we may lose the focus on the original problem (Ho, 2002). Nonetheless, MCSs become more and more popular in pattern recognition and they are capable of improving the recognition capabilities of single classifiers significantly, as it will be shown for the specific task of emotion recognition in Section 4.

3.4 Classification Setup

The first step of the classification phase is accomplished by the simple KNN classification algorithm, as shown in Figure 1. This algorithm is capable of separating data based on the assumed similarities between various feature vectors and the computed codebooks (Kohonen, 2001). In this work, the Euclidean distance between the prototypes and the presented feature vectors is considered. At this point of the classification phase a temporal fusion of the data is achieved by summing up the results for all the frames extracted from one utterance. The temporally fusioned results, which are normalized according to the number of votes per utterance, resulting in values between 0 and 1[2], of all three feature streams are then forwarded to the second part of the classification phase, the decision fusion. The decision fusion in this work is done by two simple fusion methods, which will be compared to each other in Section 4. The first one is simply summing up the output of the three different KNN classifiers and the second one is multiplying the outputs. In both cases the maximally supported class is chosen as the common classification result (see Kuncheva, 2004 for details on classifier fusion). Although it has been argued in the past which fusion technique should be preferred, a common sense has not been found, since the results strongly depend on the data. Multiplication fusion is supposed to perform better if the data streams are statistically independent and values close to zero do not occur erroneously, and summation is more stable toward statistical outliers. The performance of the single classifiers as well as various fusion techniques will be discussed in Section 4.

4. Experiments and Results

All the experiments in this work are carried out on the German emotional database, Berlin Database of Emotional Speech, which is described in Section 2. First, the data set was divided randomly, in order to conduct a 10-fold cross-validation, into 10 different data sets, containing data for testing as well as for training. A tenth of the over 400 available utterances was used as test set in every fold, the remainder was used for training. Second, the classification results of the KNN classifiers for each feature stream, as shown in Figure 1, were computed for each fold of the cross-validation.

In Table 3 the error rates in percentage of the different setups using different values for k_m indicating the number of prototypes in the codebooks for each of the seven emotions, k_n the number of nearest neighbors in the classifier, and using the two mentioned fusion techniques, are listed. These values indicate a classification accuracy of slightly over 70% in the empirically best case ($k_m = 200$ and $k_n = 1$). The error rates of this case are plotted in Figure 9, along with the classification accuracy for the two mentioned simple fusion methods. The displayed boxes indicate the range of 50% of the results of the experiments. The middle line shows the median and the whiskers indicate the range of all the classification results.

Table 3. Classification error rates (%) of fusion using different setups.

	Multiplication	Summation
$k_m = 100$, $k_n = 1$	32.55	34.43
$k_m = 100$, $k_n = 3$	31.84	32.55
$k_m = 200$, $k_n = 1$	29.95	31.84
$k_m = 200$, $k_n = 3$	30.90	30.42

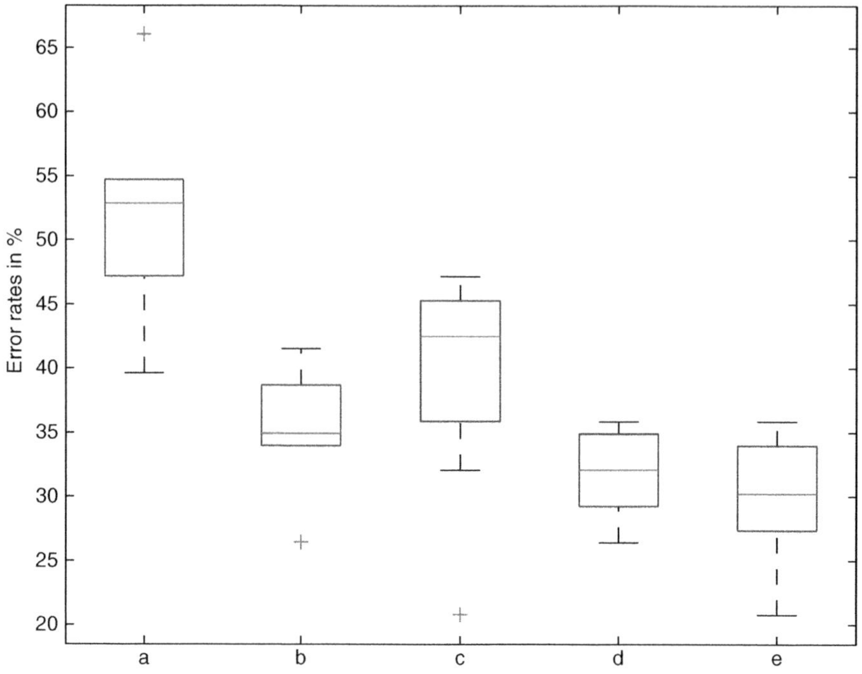

Figure 9. Comparison of the classification accuracy of the single classifiers and the decision fusion: **(a)** spectral energy modulation (8 features), **(b)** loudness after Zwicker (24 features), **(c)** RASTA-PLP coefficients (13 features), **(d)** sum fusion, and **(e)** multiplication fusion.

Outliers are displayed as crosses and are further distant from the boxes as 1.5 times the range of the corresponding whisker. It is observed that the fusion techniques (letters *d* and *e*) are superior to the single classifications in nearly all the cases. The median values of the error rates lie significantly below the ones of the single classifiers. One could argue that the importance of the modulation spectrum-based features is minor or could even reduce the accuracy of the fusion, by only looking at Figure 9. However, when removing these features the error rate is increased by around 1%. This indicates that the combined classification is capable of compensating errors made by the single classifiers. Which errors are affected in detail can be observed in Tables 4 and 5, showing the confusion matrices for the fusion classification and as an example for the single classification the results using only RASTA-PLP coefficients. In the confusion matrices the entry c_{ij} represents the number of samples from class i classified as class j.

Furthermore, the combined classifier is capable of separating agitated from calm emotions with an accuracy of nearly 95%. In Table 4 the first four emotions represent agitation as mentioned in Section 3.1 and the last three represent calm emotions. It is observed that hardly any confusions are made outside the

Table 4. Confusion matrix of the multiplication fusion experiment.

	Fear	Anger	Happiness	Disgust	Neutral	Sadness	Boredom
Fear	25	4	4	8	7	7	1
Anger	0	100	2	3	0	0	0
Happiness	3	18	26	7	0	0	0
Disgust	2	0	0	30	2	3	0
Neutral	0	0	0	1	39	1	21
Sadness	0	0	0	0	1	49	0
Boredom	0	0	1	0	20	11	28

Table 5. Confusion matrix of the classification using RASTA-PLP coefficients.

	Fear	Anger	Happiness	Disgust	Neutral	Sadness	Boredom
Fear	12	6	6	8	7	15	2
Anger	3	84	10	3	3	2	0
Happiness	6	15	27	4	1	1	0
Disgust	2	1	1	25	3	4	2
Neutral	0	0	0	2	33	8	19
Sadness	0	0	0	0	3	47	0
Boredom	0	0	0	0	16	15	29

two areas representing these groups. Only fear is confused several times with neutral and sadness. Again the multi-classifier was capable of improving the results achieved using only one set of features.

It is observed that the confusion of anger with happiness and fear with sadness could be improved significantly. However, confusions such as neutral with boredom remain nearly the same.

Emotions like neutral and boredom lie very close to each other and even for humans it is not trivial to separate them, as it is seen in the results of the human perception test listed in Table 1. Except for the confusions between happiness and neutral, disgust and sadness, and sadness and fear the two regions of agitated and calm emotions are found in the results of the human perception test as well. For humans and the automatic system the separation of boredom and neutral seems to be a hard task resulting in a large confusion in both cases. However, there are not only similarities found in the results of the multi-classifier and the human perception test. For example, it seems to be hard for humans to separate boredom and sadness, whereas the classifier succeeds in separating the two classes. Vice versa humans outperform the classifier in keeping happiness and anger apart as well as boredom and sadness.

Furthermore, it is observed from Table 4 that fear was mostly confused with disgust, happiness with anger, anger with disgust, neutral with boredom, and boredom with sadness. Some of these results are consistent with the studies by Murray and Arnott, where the relationship between emotion and speech parameters was investigated (Murray and Arnott, 1993). For example, for fear and disgust the speaking rate was much faster and pitch range wider. For anger and happiness the speaking rate was faster, the pitch average was higher, and the pitch range was much wider. Moreover, they found the articulation to be normal for happiness and disgust, tense for anger, precise for fear, and slurring for sadness (Murray and Arnott, 1993).

Scherer et al. (2003) summarize several studies and list similarities between neutral and boredom, and sadness and boredom. For boredom the fundamental frequency mean, range, and gradients are less or equal to neutral speech and jitter as well as shimmer are equal. None of the other emotions mentioned in the study by Scherer et al. have more overlap with neutral than boredom. Furthermore, for sadness and boredom the fundamental frequency is lower compared to neutral speech. However, the characteristics related to fundamental frequency of sadness are stronger as of boredom, but they lead in the same direction. It is also mentioned in Scherer et al. (2003) that sadness and boredom correspond to each other according to speech rate and fluency patterns comprising a lowered number of syllables per second, higher syllable duration, longer, and more pauses. At this point it is important to note that the findings in Murray and Arnott (1993) and Scherer et al. (2003) are not only based on German language as the database used in this study. However, European emotional expression is considerably similar over different languages.

The proposed classification fusion approach is performing better than earlier work on acted emotional English speech using large sets of features and trained ANNs, recognizing eight different emotions with an accuracy of around 50% (Nicholson et al., 2000). Additionally, the performance is comparable to the classification accuracy on Japanese in Lee et al. (2004), in which only four different emotions are considered, whereas in this work seven different emotions are targeted. The performance of the HMMs in Lee et al. (2004) is only slightly better than the one using simple fusion techniques. The HMMs recognize the four different emotions with an accuracy of 76%, whereas the multi-classifier recognizes seven emotions with 70% accuracy.

Additionally, we performed another set of experiments using only five different emotions, which are more common than others in applications where voice is the primary communication channel, such as automated call centers. We removed disgust and fear from the possible seven emotions and investigated how the performance would improve in this easier scenario. An accuracy of 76% was achieved indicating an increase of accuracy while recognizing less classes.

Moreover, the generalization ability of the system is investigated by executing leave-out one-speaker (LOOS) experiments. As mentioned in Section 2 there are 10 different speakers available in the database of emotional speech. For the experiment every speaker was used for testing once and for training in all the other nine cycles. The results of the LOOS experiment are listed in Table 6 and Figure 10. It is observed that for five speakers in the empirically best ($k_m = 100$ and $k_n = 3$) experiment the error rate did not increase significantly compared to the results shown in Table 3. However, the recognition performance did decrease for the others. Recognition performed especially well for speaker 6 of whom the least utterances are available leading to the largest set of available training data. Of speakers 9 and 10 there are the most utterances available in the database resulting in smaller training sets. This indicates that the more data available the better the generalization performance of the system.

Table 6. Classification error rates (%) of fusion using different setups for every speaker and the average.

Speaker Nr.	Multiplication		Summation	
	$k_m = 100$ $k_n = 3$	$k_m = 200$ $k_n = 3$	$k_m = 100$ $k_n = 3$	$k_m = 200$ $k_n = 3$
1	32.6	32.6	34.6	32.6
2	35.0	36.8	33.3	38.5
3	34.8	37.2	34.8	37.2
4	45.9	54.1	48.6	54.1
5	43.6	50.9	43.6	54.5
6	24.2	27.2	24.2	24.2
7	39.3	34.4	39.3	36.0
8	33.8	29.4	32.3	25.0
9	48.2	44.6	50.0	48.2
10	47.8	53.5	52.1	53.5
Avg.	38.5	40.0	39.3	40.0

Again, in Figure 10 it is observed that the fusion techniques (letters *d* and *e*) are outperforming the single classifiers, although the difference is not as clear as in Figure 9. The superior classification performance is pointed out by the red lines within the boxes, which mark the median values. Additionally, in Figure 10 it is noticed that the scatter of the results is much larger than in the 10-fold cross-validation experiments. This may be the case, because the experiments are highly dependent on the acting quality of the actor that is being tested. Some perform in a very special and unique way, which cannot be captured by the automatic emotion recognizer. However, in half of the cases the

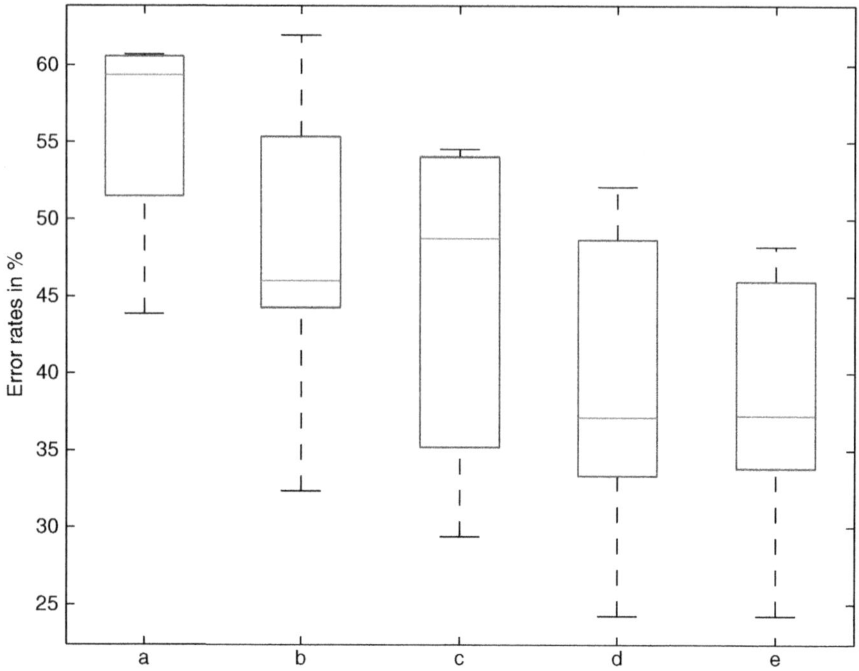

Figure 10. Comparison of the classification accuracy of the single classifiers and the decision fusion in the leave-out one-speaker experiment. **(a)** spectral energy modulation, **(b)** loudness after Zwicker, **(c)** RASTA-PLP coefficients, **(d)** sum fusion, and **(e)** multiplication fusion.

generalization ability is very strong and the system is capable of recognizing emotions accurately.

5. Conclusion

The work presented in this chapter introduced three uncommon features and a simple yet accurate multi-classifier system for emotion recognition. The utilized features comprise RASTA-PLP, perceived loudness, and modulation spectrum features. All the features are inspired by the human auditory system. The usefulness of the mentioned feature sets is shown in several experiments, such as the classification of agitated and calm emotions and the combined experiments to classify up to seven different emotions. In order to handle the huge amount of data extracted from the emotional utterances a data reduction step was implemented. The common and simple k-means algorithm was used to reduce the data to codebooks comprising representative prototypes. Additionally, the benefit of multi-classifier systems is discussed and shown in our special case conducting classification experiments with single and multi-classifiers. Additionally, two fusion methods were compared resulting in similar recognition performances. The fusion of the simple KNN classifiers for each of the three feature sets lead to improved results throughout all the experiments. Clear improvements were achieved without increasing the required computing time too much, since only simple fusion methods like the summation or multiplication of the single classifier outputs were used.

The performance of the proposed system is comparable to earlier work where large sets of features were required to recognize the emotions with complex trained ANNs or HMMs. Additionally, the performance of the automatic emotion recognizers was compared to human performance in a perception test on the same standard database. The results have indicated similarities between the confusion matrices of the human perception test and the automatic recognizers. Furthermore, similarities between earlier speech analysis and the classification results of the proposed approach could be found. Additionally, the system showed its ability to deal with unforeseen conditions, such as unknown speakers, in a leave-out one-speaker experiment. The recognition performance did not drop drastically for most of the speakers.

However, only simple classifiers and simple fusion methods are used and could be extended toward trained fusion techniques to further improve the performance of the automatic emotion recognizer. Additionally, the proposed features could be used along with the standard approaches such as pitch and its variations. Experiments using different microphones, telephone quality recordings, and additional noise could prove the robustness of the proposed features and multi-classifier system. These issues will be studied in the future.

Acknowledgments This work is supported by the competence center Perception and Interactive Technologies (PIT) in the scope of the Landesforschungsschwerpunkt project: "Der Computer als Dialogpartner: Perception and Interaction in Multi-User Environments" funded by the Ministry of Science, Research and the Arts of Baden-Württemberg.

Notes

1. "The black sheet of paper is there besides the piece of wood."
2. These values can be interpreted as probabilities/support of a single class by the KNN classifier.

References

Burkhardt, F., Paeschke, A., Rolfes, M., Sendlmeier, W., and Weiss, B. (2005). A Database of German Emotional Speech. In *Proceedings of Interspeech.*

Cowie, R., Douglas-Cowie, E., Tsapatsoulis, N., Votsis, G., Kollias, S., Fellenz, W., and Taylor, J. (2001). Emotion Recognition in Human-Computer Interaction. *IEEE Signal Processing Magazine*, 18:32–80.

Dellaert, F., Polzin, T., and Waibel, A. (1996). Recognizing Emotion in Speech. In *Proceedings of International Conference on Spoken Language Processing (ICSLP1996)*.

Devillers, L., Vidrascu, L., and Lamel, L. (2005). Challenges in Real-Life Emotion Annotation and Machine Learning based Detection. *Neural Networks*, 18:407–422.

Fragopanagos, N. and Taylor, J. (2005). Emotion Recognition in Human-Computer Interaction. *Neural Networks*, 18:389–405.

Hermansky, H. (1996). Auditory Modeling in Automatic Recognition of Speech. In *Proceedings of Keele Workshop*.

Hermansky, H. and Morgan, N. (1994). Rasta Processing of Speech. *IEEE Transactions on Speech and Audio Processing, Special Issue on Robust Speech Recognition*, 2(4):578–589.

Hermansky, H., Morgan, N., Bayya, A., and Kohn, P. (1991). Rasta-PLP Speech Analysis. Technical report, ICSI Technical Report TR-91-069.

Ho, T. K. (2002). Multiple Classifier Combination: Lessons and Next Steps. *Series in Machine Perception and Artificial Intelligence*, 47:171–198.

Kanederaa, N., Araib, T., Hermansky, H., and Pavele, M. (1999). On the Relative Importance of Various Components of the Modulation Spectrum for Automatic Speech Recognition. *Speech Communications*, 28:43–55.

Kohonen, T. (2001). *Self Organizing Maps*. Springer, New York.

Krishna, H. K., Scherer, S., and Palm, G. (2007). A Novel Feature for Emotion Recognition in Voice Based Applications. In *Proceedings of the Second International Conference on Affective Computing and Intelligent Interaction (ACII2007)*, pages 710–711.

Kuncheva, L. (2004). *Combining Pattern Classifiers: Methods and Algorithms*. Wiley, New York.

Lee, C. M., Yildirim, S., Bulut, M., Kazemzadeh, A., Busso, C., Deng, Z., Lee, S., and Narayanan, S. S. (2004). Emotion Recognition Based on Phoneme Classes. In *Proceedings of International Conference on Spoken Language Processing (ICSLP2004)*.

Murray, I. R. and Arnott, J. L. (1993). Toward the Simulation of Emotion in Synthetic Speech: A Review of the Literature on Human Vocal Emotion. *Journal of the Acoustical Society of America*, 93(2):1097–1108.

Nicholson, J., Takahashi, K., and Nakatsu, R. (2000). Emotion Recognition in Speech Using Neural Networks. *Neural Computing and Applications*, 9(4): 290–296.

Oudeyer, P.-Y. (2003). The Production and Recognition of Emotions in Speech: Features and Algorithms. *International Journal of Human Computer Interaction*, 59(1–2):157–183.

Petrushin, V. (1999). Emotion in Speech: Recognition and Application to Call Centers. In *Proceedings of Artificial Neural Networks in Engineering*.

Rabiner, L. R. and Schafer, R. W. (1978). *Digital Processing of Speech Signals*. Prentice-Hall Signal Processing Series. Prentice-Hall, Upper Saddle River.

Scherer, K. R., Johnstone, T., and Klasmeyer, G. (2003). *Handbook of Affective Sciences – Vocal Expression of Emotion*, Chapter 23, pages 433–456.

Scherer, S., Schwenker, F., and G., P. (2008). *Emotion Recognition from Speech Using Multi-Classifier Systems and RBF-Ensembles*, Chapter 3, pages 49–70.

Whissel, C. (1989). *The Dictionary of Affect in Language*, Volume 4 of *Emotion: Theory, Research and Experience*. Academic Press, New York.

Yacoub, S., Simske, S., Lin, X., and Burns, J. (2003). Recognition of Emotions in Interactive Voice Response Systems. In *Proceedings of the European Conference on Speech Communication and Technology (Eurospeech)*.

Zwicker, E., Fastl, H., Widmann, U., Kurakata, K., Kuwano, S., and Namba, S. (1991). Program for Calculating Loudness According to DIN 45631 (ISO 532B). *Journal of the Acoustical Society of Japan*, 12(1):39–42.

Chapter 6

UNDERSTANDING MOBILE SPATIAL INTERACTION IN URBAN ENVIRONMENTS

Katharine S. Willis
MEDIACITY Project, Bauhaus University of Weimar, Weimar, Germany
katharine.willis@archit.uni-weimar.de

Christoph Hölscher, Gregor Wilbertz
Centre for Cognitive Science, University of Freiburg, Freiburg, Germany
{hoelsch, gregor.wilbertz}@cognition.uni-freiburg.de

Abstract In order to act in urban environments an individual accesses various types of knowledge, such as memories, spatial strategies and also information from the environment so as to develop plans and make decisions. This chapter will investigate the nature of spatial knowledge acquisition in an environmental setting by comparing performance in a task where the participants learnt the environment using spatial assistance either from a map or from a mobile map. It outlines the early results of an empirical experiment which evaluated participants' spatial knowledge acquisition for orientation and distance estimation tasks in a large-scale urban environmental setting. The initial findings of the experiment highlight the fact that mobile map participants performed worse in distance estimation tasks than map participants and that their errors for complex routes were high. We will conclude by analysing the results of this experiment in terms of the specific types of knowledge afforded by mobile maps and the implications for spatial learning in urban environments.

Keywords: Wayfinding; Spatial knowledge; Mobile maps; Maps.

1. Introduction

We learn about the spatial environment through a process referred to as cognitive mapping, which is a series of psychological transformations by which

W. Minker et al. (eds.), *Advanced Intelligent Environments*,
DOI 10.1007/978-0-387-76485-6_6,

an individual acquires, codes, recalls and decodes information about relative locations and attributes in a spatial environment (Downs and Stea, 1974). It is a dynamic process and occurs following successive experiences of sequential routes, where knowledge about the environment is integrated into configurational survey representations (Siegel and White, 1975). At a global level we use survey-type knowledge that distinguishes between here and there and regions around and between them. We use this information to determine a plan for travelling between one place and another. Once we set out on a path, we need local representations of our surroundings to help us make decisions at key points. All these types of knowledge are used to try to prevent us from losing our way and to ensure we arrive at our desired destination. However, we also use spatial assistance of many kinds to augment the knowledge in our heads. This includes maps and also more recently developed mobile maps of navigation assistance supported by GPS. Mobile maps seem to provide an ideal solution to the problem of getting local information on where to go whilst completing a wayfinding task; this information is incremental since it is delivered in stages, rather than a paper map source which provides all the information in a stable format and is usually studied primarily in the planning stage of a task. However, recent empirical studies found that participants who used mobile maps to navigate an environment had significantly poorer survey knowledge acquisition when compared to participants who used paper maps in the same task (Aslan et al., 2006). In order to investigate this further, we will first analyse how people acquire knowledge about urban space and the role of representations in this process. In order to better understand the nature of spatial knowledge acquisition with mobile maps we then describe an empirical experiment which evaluated spatial knowledge acquisition in a large-scale environmental setting by comparing participants who had learned the environment from a map and participants who had learned it using a mobile map.

2. Approach and Hypothesis

2.1 Spatial Knowledge Acquisition in Urban Environments

In a large-scale environment the structure and features are revealed by integrating local observations over time, rather than being perceived from one vantage point (Kuipers, 1982). This means that the acquisition of spatial knowledge of urban space depends not just on direct perception but on learning experience. One of the key elements of this experience of an environment in motion is the paths along which we move. People observe the environment whilst moving through it on the path, and environmental elements are perceived as arranged and related along these paths (Lynch, 1960). The behavioural pattern or movement performed by moving along a path is called a route. When moving along a path an individual experiences an organised sequence in which phases follow each other meaningfully in an order. Consequently studies (Golledge et al., 1995; Hirtle and Hudson, 1995; Peponis et al., 2004) have found that the route sequence affects learning. For instance, in the Golledge study, participants who undertook a wayfinding task within a very simple regular environment performed significantly worse when they took one route than when they took another, even when the routes were very similar on the plan. Thus the difference in the way the environment was experienced as a certain sequence of events and features significantly affected the degree of error with which the individual completed the task. This can be illustrated simply with an example we have probably all experienced. A route or path which when walked in one direction is familiar and legible, but often completely unrecognisable when walked in the opposite direction. It appears that the difference in the dynamic or sequential experience of the environment, and therefore its configuration, affects comprehension.

2.2 Knowledge from Maps and Mobile Maps

Studies of spatial knowledge acquired through maps and through navigation have found that when people learned the environment from a map they had different knowledge about it, compared to those who had experienced it only through navigation (Thorndyke and Hayes-Roth, 1982). Map users' acquire a survey knowledge of the environment whereas individuals who have acquired knowledge through navigation attain a configurational knowledge where features and structure have been integrated over time (Siegel and White, 1975). In using a map a person must construct an adequate mental representation of the one-to-one mapping between the space he/she experiences from an egocentric viewpoint and the bird's eye view of the environment offered by the map. So far few studies have looked at the knowledge acquired from mobile maps, which present incremental information during navigation of an environment. However, we propose here that mobile maps support the acquisition of route-type knowledge, because the individual learns the environment in a procedural egocentric manner. Route knowledge is structured around decision points along the path, which are remembered in a sequential and 'chunked' manner, and the path is characterized by the location of landmarks along it. There are many open questions, however, as to whether the dynamic presentation of route directions, such as those in a mobile map, supports spatial knowledge acquisition or whether it merely succeeds in guiding an individual from a start point to a destination, but without supporting the spatial learning process.

2.3 Perception and Memory in Spatial Tasks

A good deal of the time, we proceed in wayfinding tasks based on piecemeal spatial information in unfamiliar or semi-familiar environments where we do not have access to learned configurational structures. In these cases we often adopt strategies based on our previous experience of similar environments; this is a form of hard-wired background knowledge (Portugali, 1996; Raubal et al., 1997) or 'image schemata' (Johnson, 1987; Neisser, 1976), which by means of imagination are used to help us relate to the world and categorize it. These schemata are the recurrent mental patterns that help people structure space. For example when we approach a new city for the first time and arrive at the central station, we will attempt to formulate a travel plan based on our experience of other cities and the various possible positions of the central station. We will make assumptions about the structure and features of the city and mentally update our schematic framework, and as we move through the city we encounter information that either confirms or rejects our suppositions. In addition to the schemata there are 'cues' (Lynch, 1960; Couclelis et al., 1987) or salient features of the environment, which tend to be located at decision points and comprise Lynchian structures such as landmarks, nodes, paths, edges and districts. These features need not necessarily have strong visible features or shape, as one might expect from a typical landmark, instead what is important is the way in which they work in conjunction with the organisation of the space, and thus the degree to which they are perceived as salient or legible as the individual moves through this environment. However, legibility is also intertwined with the way an individual perceives the world when one studies task-based activities, such as finding one's way from A to B, where perception and memory process occur simultaneously. In these situations, environments are perceptually complex and always provide more information than can possibly be processed.

3. Learning from Field Studies

In order to understand how individuals acquire spatial knowledge about the environment when they use mobile devices, a field study was undertaken which replicated critical features of the seminal study by Thorndyke and Hayes-Roth (1982). The Thorndyke and Hayes Roth (hereafter abbreviated to T&HR) study evaluated spatial knowledge acquisition by comparing participants who had navigation experience, with those who had experienced the same environmental setting only by learning the map. The analysis found that map participants made good estimations of direct distances between two points, whereas navigation participants had better orientation abilities and performed well in route distance estimations. This chapter presents early results from an experiment in which we compared maps and mobile maps, thus varying and extending the original comparison of maps and pure navigational experience. The experiment also differed from the original study in that it took place in an external environmental setting, rather than a building. However, the structure, layout, location of features and scale of the setting were directly comparable to that of the original T&HR study.

3.1 Experiment Method

The environmental setting chosen was an allotment area located near the University of Bremen Campus, Germany, approximately 400 m $\times$ 300 m in size. The setting is external and in an urban area, although somewhat removed from a standard urban environment. It comprises a series of interconnected paths (streets), numerous small houses with garden plots laid out in rows, which are interspersed with a few larger buildings (see Figure 1).

Figure 1. Aerial view of the experiment environment.

In terms of environmental features and structure it recreates a typical urban residential area, although on a slightly smaller scale. The setting layout comprises a rectilinear grid path structure (indicated by white blocks with dotted lines in Figure 2), with several prominent visual landmarks (indicated by black rectangles).

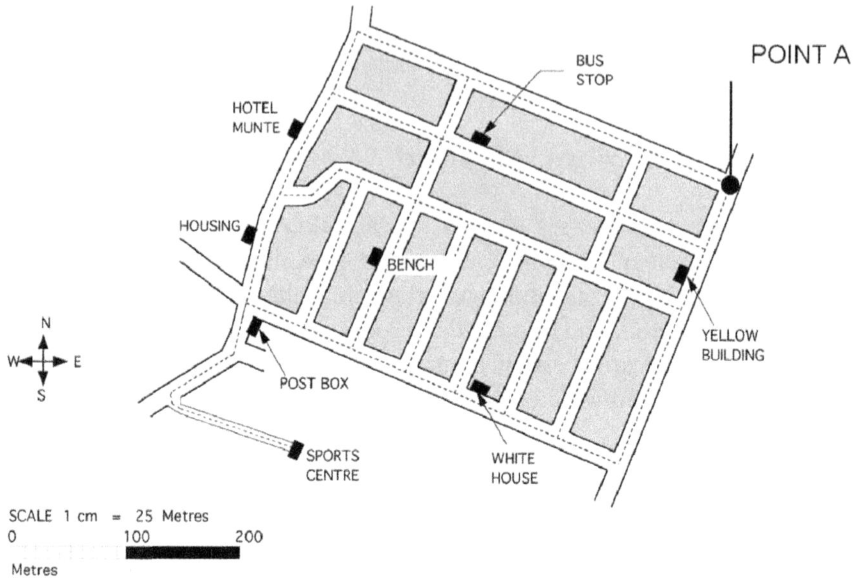

Figure 2. Map of the environment used in the map condition.

There are no visual clues or landmarks that enable global orientation, and most landmarks were not visible from one location to another.

3.2 Participants

Twenty-four participants participated for an honorarium of 7.50 Euro per hour. There were 12 female and 12 male participants. Participants had a range of academic and non-academic backgrounds and ages.

3.3 Experiment Design

The experiment comprised two training conditions with approximately 12 participants per condition. None of the participants had any prior experience of the environmental setting. In the first condition, map participants were seated in an office without visual connection to the environmental setting and asked to learn a map of the environmental setting. In the second condition, mobile map participants learnt the map through navigation with the mobile device in the environmental setting itself. Once both groups of participants had completed the learning task they were asked to complete a series of estimation tasks in the environment. For each item the participants performed three estimates: orientation (pointing to the start point in the direction of the destination), Euclidean distance (the straight line distance from start point to destination) and route distance (the shortest distance from start point to destination along the paths).

3.4 Procedure

Participants were tested individually. Each map-learning participant was told that he/she was to learn the map of the environmental setting (shown in Figure 2), including the layout of the paths, and the names and locations of the landmarks. Each participant studied the map on a series of study-recall trials, where the participants were given the map to study for 2 minutes and were then asked to redraw the features of the map on a sheet of A4 paper. This was repeated until the participant had reproduced the topological properties of the map and labelled them correctly.

In the mobile map condition the participants learnt the map in an environmental setting. The map was presented on a Nokia 6630 mobile phone (see Figures 3 and 4) running a mobile mapping software application. A GPS tracking device indicated the participants' real-time location on the map with a moving dot.

Once they had learnt the environment, the participants were asked to complete the same study-recall task as completed by the map participants, but completed this task in the environmental setting. If after the first attempt to recall and draw the map features the mobile map participant was unsuccessful, then

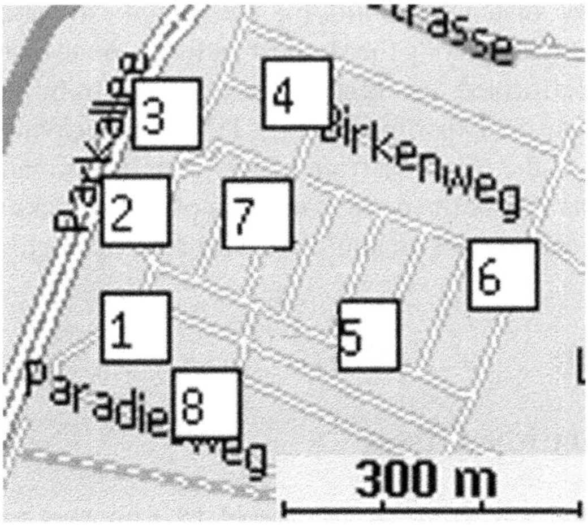

Figure 3. Screenshot of the map of the environment used in the mobile map condition.

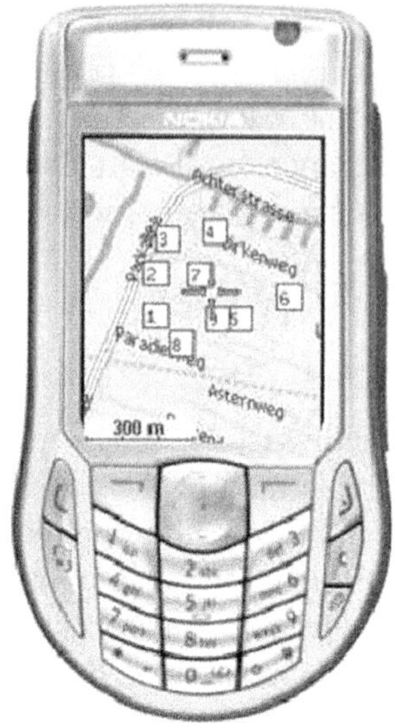

Figure 4. Screenshot of the map of the environment used in the mobile map condition.

he/she was given the mobile map device in a static condition to study for 2 more minutes.

For the estimation task, the experimenter took each participant to the first start point, the Post Box. The experimenter then informed the participant that she would read to him/her a succession of locations within the environmental setting in a random order. For each location, the participant was to perform three estimates. First, the participant indicated to the nearest degree the direction to the destination, using the compass (i.e. the orientation). Second, the participant estimated the distance in metres to the destination as the crow flies (i.e. the Euclidean distance). Third, the participant estimated the distance in metres to the destination along the shortest path (i.e. route distance). When the participant had performed the eight sets of estimates from the first start point from the Post Box, the experimenter then led the participant to the next start point. The order of the positions from which participants made their estimates was varied, so that half of the participants completed a clockwise route and the other half of the participants an anti-clockwise route in the test phase.

4. Result

4.1 Learning Phase

The map participants took an average of 18.3 minutes to learn the paper map in the office setting. The mobile map participants, who undertook the learning task in the environment, spent an average of 46.8 minutes learning the mobile map in the environment (although the time each participant spent interacting with the mobile map varied). The maps drawn by the participants in the learning phase were studied for schematic differences. There were qualitative differences between the features, and the quality of the maps depended on whether the participant had learned the environment using the map or the mobile map. Consequently, it appears that the basic quality and features of knowledge acquired by the two groups in the learning task are different.

The mobile map participant (for example, see Figure 5) typically indicated a route, with a series of sequential landmarks. In this case the configuration of the environment is not shown, and no features were indicated which were not on the route. None of the mobile map participants redrew the map with the layout facing true 'north-up' despite the fact that they had accessed a map representation which was orientated 'north up' throughout the learning task. In addition all the participants rotated the map so that the paths were orientated parallel to the x, y axis whereas the paths on the map are orientated at an angle of approximately 20° to the y axis – see screenshot in Figure 3.

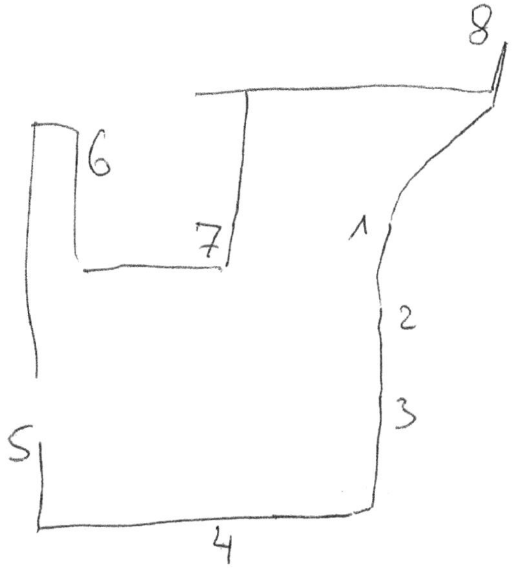

Figure 5. Example of first map drawn by the mobile map participant in learning task.

In contrast the map participants (for example, see Figure 6) typically redrew the map to depict a clear survey structure onto which landmarks and features are placed. All map participants redrew the map 'north-up'. This suggests that the map participants very quickly assimilated a strong and stable image of the environment in the learning task.

Figure 6. Example of first map drawn by the map participant throughout the learning task.

In contrast the map participants (for example see Figure 6) typically redrew the map to depict a clear survey structure onto which landmarks and features are placed. All map participants redrew the map 'north-up'. This suggests that the map participants very quickly assimilated a strong and stable image of the environment in the learning task.

4.2 Orientation Estimates

We tested the predictions for the orientation task by contrasting the average angular error between the true and the estimated orientations of destinations for the map and mobile map participants. Generally the performance was very good, with the difference between real and estimated distances being generally small. The average angle of error for map participants was $16.31°$, and for mobile map participants it was $16.58°$ – see Table 1.

In this case errors were not symmetrically distributed around the average value, with the majority of the participants having lower value differences (map $11°$, mobile map $14°$). There was no difference in the systematic distribution (with the distribution to the right for map being 53% and mobile map 51%). In the median there was a slight tendency for the differences to be lower for map participants, although this difference was not significant, whether tested

Table 1. Angular error (degrees) for orientation judgements.

Type of experience	Angle error
Map	16.31
Mobile map	16.58
Thorndyke and Hayes Roth (T&HR) map	40
T&HR navigation	19

parametrically (t-test) or non-parametrically (Mann–Whitney test). Critically, there was no effect on the results of the destination being visible to the participant, compared to when the destination was not visible.

4.3 Euclidean Estimates

The difference between the estimated and actual distances is large (map: 69.73 m, mobile map: 94.33 m) – see Table 2.

Table 2. Relative error (percent) in Euclidean estimates.

Type of experience	Euclidean distance error
Map	33
Mobile map	40
T&HR map	32.8
T&HR navigation	32.3

There is also indication of an uneven distribution around the median (map: 67.89, mobile map: 75.90). Both groups underestimated the actual distance more frequently (map 65%, mobile map 73%) than they overestimated (correspondingly 35%, 27%). The systematic differences averaged with the map were -12.43, and mobile map -22.99. Between the two groups the relative differences were compared and found to be not significant (map: 0.33, mobile map: 0.40, $p = 0.222, d = 0.52$). Looking at individual tasks separately, only 2 of 42 tasks showed significant differences (largely due to high inter-individual variance). Yet in the majority of the tasks the map users performed better (with an average effect size of $d = 0.46$), whilst the mobile map users were only better with one-quarter of the tasks and for these the differences are much less pronounced (average effect size: $d = 0.20$). The correlation between estimated and actual differences is equally high in both the groups (map: $r = 0.78$, mobile map: $r = 0.79$). This pattern is independent of the testing method, whether parametric or non-parametric, and also independent of whether the destination was visible from the start point.

4.4 Route Distance Estimates

The difference between the estimated and actual distances is larger than that for the Euclidean distance (map: 98.39 m, mobile map: 135.90 m). There is also indication of an uneven distribution around the median (map: 77, mobile map: 99). Both groups underestimated the actual distance more frequently (map 71%, mobile map 72%) than they overestimated (correspondingly 29%, 28%). The average of the systematic differences was for the map -43.55 and for mobile map -50.27. The relative differences were tested between the groups: the difference (with individual tests) is marginally significant (map: 0.32, mobile map: 0.42, $t(12.88) = -20; p = 0.067, d = 0.82$) – see Table 3.

Table 3. Relative error (percent) in route distance estimates.

Type of experience	Route distance error
Map	32
Mobile map	42
T&HR map	35.8
T&HR navigation	26.2

However, the non-parametric test was not significant. Looking at individual tasks separately, we observe statistically significant differences for only 2 of 42 tasks. But again a clear tendency becomes visible: in 90% of the tasks the map users numerically outperform the mobile map users, confirming the parametric test above.

4.5 Comparison Between Distance Estimates

The difference between Euclidean and route distance estimates is quite small and provides a statistical trend only within the group of mobile map participants (0.40 vs. 0.42, $t(11) = -1.91, p = 0.083$) – see Table 4.

Table 4. Relative error (percent) in distance estimates.

Type of experience	Euclidean error	Route distance error
Map	33	32
Mobile map	40	42

4.6 Dependence on Number of Legs

Generally the relative deviation increases with increase in number of legs - see Tables 5 and 6.

Table 5. Relative error (percent) in Euclidean estimates, showing dependence on no. of legs.

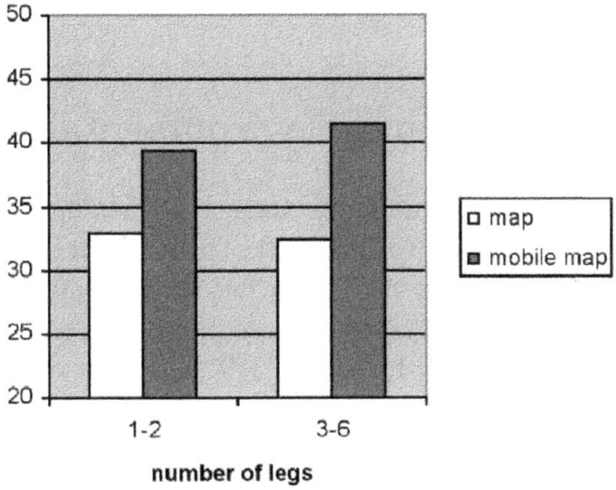

Table 6. Relative error (percent) in route distance estimates, showing dependence on no. of legs.

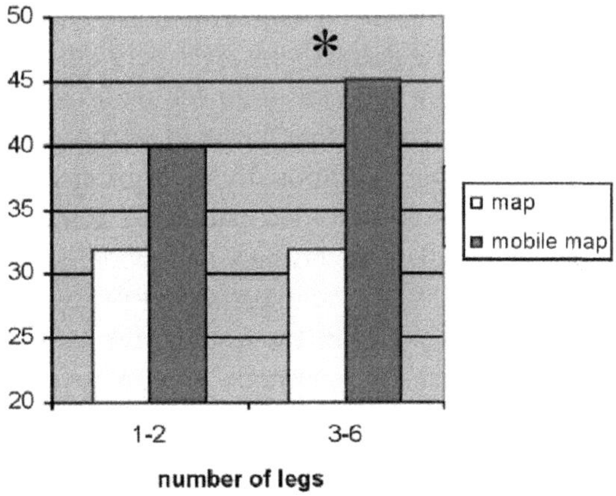

For map participants there did not appear to be any effect of the route complexity (number of legs) on the accuracy of the estimate (distribution approximately 32–33%, left). For more complex tasks (with three or more legs) the difference between Euclidean (distribution 42%, SD=21) and route distance estimates (45%, SD=20) was significant ($p = 0.075$, df=11) for map participants. Mobile map participants generally made larger estimation errors than map users, and they made greater errors in route distance estimates when compared with Euclidean distance estimates. For the estimates on complex routes, the differences for mobile map participants (45%, SD=20) are significantly greater than for map participants (32%, SD=8; $t(14.47) = -2.00$, $p = 0.046$). This effect can be classified as large ($d = 0.89$).

5. Discussion

5.1 Orientation and Environmental Regularity

Map and mobile map participants seemed to acquire a similar level of knowledge of the orientation of landmarks in the environment, which when compared to the original study by T&HR has results similar to the navigation participants. This could be seen as a consequence of the benefit of the environmental setting. The setting for the T&HR study was one floor of an office building where participants were required to make estimates whilst in a closed room, i.e. they were isolated from any visual cues about the environment for which they were estimating. It seems that having visual access to the environment to match the mental representation and the real setting is very useful for making estimations. This is despite the fact that there was no effect of the visibility of the destination. Both map and mobile map participants registered salient visual cues in the environment and used these to match against salient aspects of their internal representation in order to make orientation estimations. In this study the map users orientated themselves very quickly at the beginning of the task and used environmental cues to match their internal map to the environment structure, despite the fact that they had learned the map without any prior experience of the setting. Additionally both groups repeatedly reported problems in making the estimation to destination D2 Bench, since it was on a path which was not part of either the primary survey or the route configuration of the environment, and thus was often spatially 'lost' by the participants. In general the regularity of the environment would have an affect on the overall success in completing the task. In extremely regular environments, i.e. those with a clear grid structure, it is easier to acquire survey knowledge. In a more irregular environment this process might be much more error prone.

5.2 Types of Knowledge

The accuracy of the metric knowledge between the two groups was different, with mobile map users making more pronounced errors. At first these results seem strange; how can an individual acquire knowledge about spatial relations but have such poor metric knowledge? The orientation data would suggest that mobile map participants had acquired a similar level of configurational knowledge about the environment as the map users. However, the analysis of the Euclidean and route distance estimates disproves this. Map learners acquired a bird's eye view of the environment which enables them, despite a very short period of learning, to demonstrate good spatial knowledge acquisition. The map participants in a sense mentally envisaged a layer of their internal map laid or stretched over the environment and used this to frame their estimations. The attention of the mobile map participants during the learning task was on a series of landmarks, which they learned along a route. It enabled the mobile map participants to identify and estimate the eight main destinations, but when they were required to make estimations of Euclidean or route distances with many legs, the performance declined more than that of map participants, indicating that the route knowledge was confined to the route they had navigated in the learning task. Conversely map participants' performance did not change with route complexity with either Euclidean or route distance, indicating the broader, configurational nature of their knowledge. When we compare the results back to the original T&HR experiment it seems fair to assume that navigation participants would have better route distance estimation than map participants. From this experiment it would be suggested that navigation participants would perform best, followed by map participants and then mobile map in route distance estimation. For Euclidean estimates, map and navigation participants were fairly similar, but mobile map participants performed worse.

Although the map representation learned by the two groups had similar features, it is clear that the way it was presented and delivered to the participants fundamentally affected the spatial knowledge acquired. For map participants, the learned knowledge acted like a configurational framework, which enabled the individual to locate landmarks within the schema (see Figure 7) and make fairly consistent orientation and distance estimates for all destinations.

Mobile map participants acquired a schema based on procedural route knowledge structured into 'chunks' or sections of learned route centred on the salient landmarks (see Figure 8).

This enabled them to make good estimates of spatial relations, but was fragmented and sequential in nature. This would explain why mobile map participants made greater errors with routes with greater complexity, because these legs would tend to lie outside the main route knowledge of the participants.

The schema frameworks used by the two groups to make estimates was therefore generally different, with mobile map users trying to reconfigure sequential route sections structured around landmark cues, and map users attempting to match their landmarks within a structured template-like schemata, which was mentally mapped onto the environment through rotation of the image.

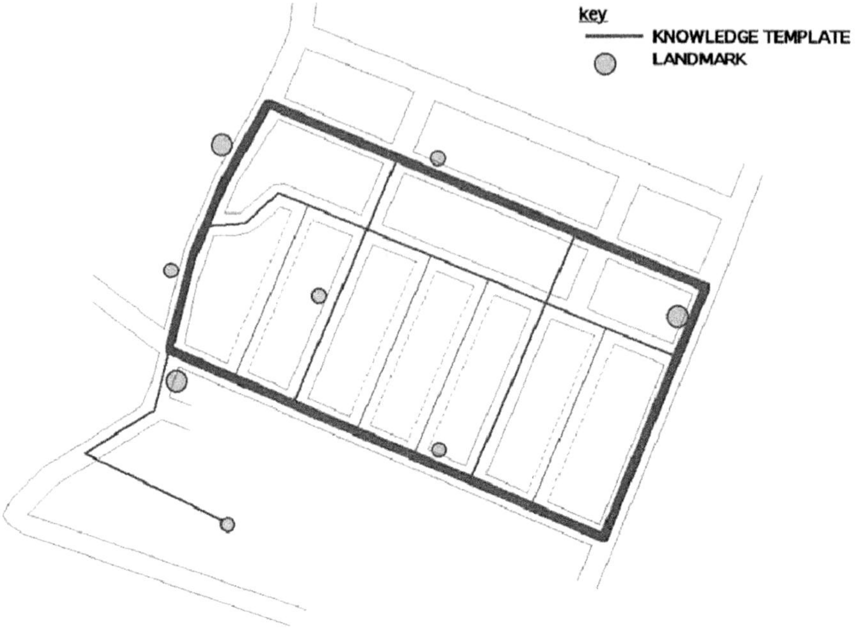

Figure 7. Map knowledge schemata.

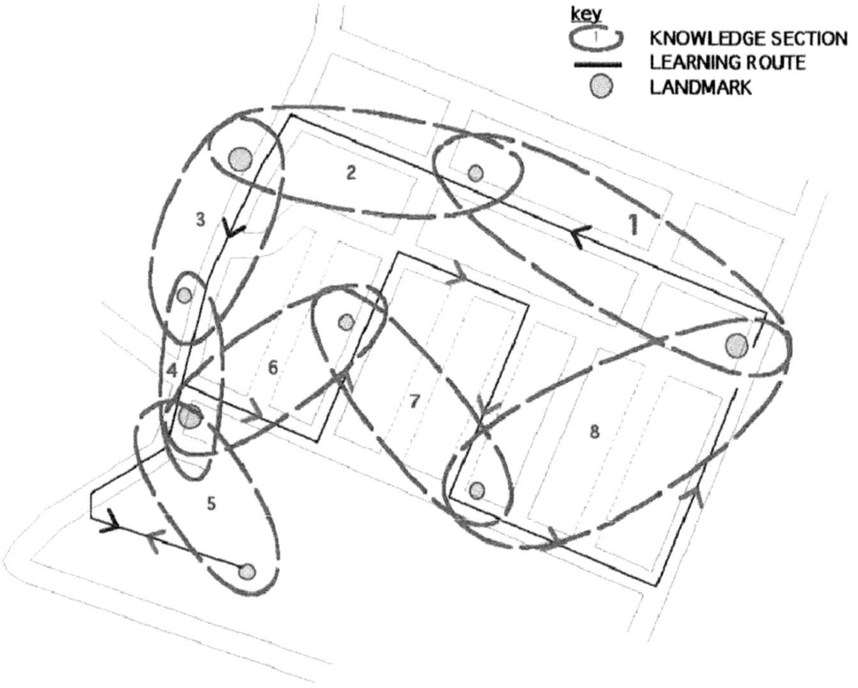

Figure 8. Mobile map knowledge schemata.

5.3 Attention

Mobile maps require attention from the individual whilst he/she is in the environment because the information is both automatically changing and updating. This is also because the individual can interact with the information (e.g. by changing the scale of the map interface). Interestingly, it is not the graphic representation of the mobile map that distinguishes it from a cartographic map in the effect on performance, but rather the delivery and attention it requires of the individual. For example a study by Gaerling and Lindberg found that participants who were engaged in a secondary task whilst navigating a route were less able to keep track of the learned locations than those groups who did not have a concurrent task (Gaerling and Lindberg, 1983). With a mobile map the fact that it constantly updates the users' position on the map itself, thus effectively offering a dynamic, constantly changing set of information, seems to create a very sequential form of spatial knowledge. This meant that their attention was being divided between the device and the environmental setting, which affected their memory. In addition since the user had the ability to change the scale of the map, meaning that they did not consistently view the same map representation throughout the task, this created a type of focused 'keyhole' information acquisition. Rather than the mobile map interface disappearing into the background it seems to have the opposite affect: creating a conflict, with attention to the features of the real environment being divided and shared with attention to the interface. Since both offered different forms of information – the mobile map a survey-type representation and the environment an egocentric perspective – the participants were in a constant process of trying to resolve the information from the map and the information from their view environment, and this was cognitively demanding. The map participants also had to translate the two perspectives to make their judgements, but paradoxically the learned map representation did not demand their attention because since it offered a single static clear representation it enabled more cognitive offloading where the individual could focus on matching cues in the environment to his/her internal map in a more cognitively adequate manner.

6. Interacting and Learning with Mobile Devices in Urban Environments

If the city is the way we perceive it and if it is formed by our memories of it, then the city is different when perceived with mobile maps, through navigation or through the use of cartographic paper maps. The knowledge acquired through mobile maps is route-like in nature. This suggests that survey knowledge has not been acquired, with the consequence that qualities such as linkage, connectivity and scale would be missing. We can relate this work back to that of Kevin Lynch where he identified a series of key elements in urban space that help individuals to structure their knowledge: landmarks, nodes, edges, districts and paths structure our understanding of urban space. Following field studies in Venezuela, Appleyard further clarified these elements as being either sequential or spatial in nature (Appleyard, 1970). Due to the strong sequential nature of learning with mobile maps, this would imply that people who used this type of spatial assistance will have less-developed concepts of spatial elements such as edges and districts and, instead, sequential elements such as landmark and path learning will be dominant. Additionally they will have more incremental knowledge, where sections of learned configurations are placed together in the sequence in which they were learned. But if we work under the assumption that our knowledge about the city is never independent of the way it is perceived and remembered, then the city encountered through such mobile mapping support will be perceived as having different structural characteristics and in many cases individuals will have poorer memory for environments where they have used mobile map assistance to find their way.

7. Conclusion and Future Work

When we move and act in the urban environment in a goal-directed manner, we acquire knowledge about it. This knowledge is transformed into mental representations. The representations can be retrieved to make decisions during navigation, but we also use graphic representations such as maps and mobile maps to assist us. In this chapter we introduced an experiment which looked at the types of knowledge acquired by an individual, depending on whether they used a map or mobile map to assist them. The study found that mobile map users performed worse than map users, particularly on Euclidean and route distance estimation, and that this was a consequence of the format and presentation of the spatial information. We examined the reasons for these differences and stated that mobile map users acquire a fragmented set of knowledge about spatial features, whereas map users act on a framework within which features are located. We concluded that this has implications for how people will perceive and act in urban environments, and that they may develop less well-formed spatial representations as a result of the type of spatial knowledge they have acquired from the mobile map assistance.

This chapter presented the early results and analysis of an empirical study. Further work is planned in order to provide further analysis of the experimental results. This will include an investigation of the effect of both route sequences in the learning phase of the experiment. Additionally a study will be carried out that will reconstruct participant's cognitive maps of the various locations using combinations of estimates for each location. This will include tests for consistency for different destination estimates, to ascertain whether the position of the destination within the environmental configuration has any effect on performance. It is also intended to develop the outcomes of the experiment into a series of suggestions for how the design of mobile map interfaces could be improved to better support spatial knowledge acquisition in urban space. These will be tested with comparative field studies in the original experiment environment.

References

Appleyard, B. (1970). Styles and Methods of Structuring a City. *Environment and Behaviour*, 28(4):471–493.

Aslan, I., Schwalm, M., Baus, J., Krüger, A., and Schwartz, T. (2006). Acquisition of Spatial Knowledge in Location Aware Mobile Pedestrian Navigation Systems. In *Proceedings of Mobile HCI '06*, Volume 159, pages 105–108. ACM, New York, USA.

Couclelis, H., Golledge, R., Gale, N., and Tobler, W. (1987). Exploring the Anchorpoint Hypothesis of Spatial Cognition. In Gärling, T., editor, *Readings in Environmental Psychology: Urban Cognition*, pages 37–57. Academic Press, New York, London.

Downs, M. and Stea, D., editors (1974). *Image and Environment, Cognitive Mapping and Spatial Behavior.* Aldine Publishing, Chicago, USA.

Golledge, R. G., Dougherty, V., and Bell, S. (1995). Acquiring Spatial Knowledge: Survey Versus Route-Based Knowledge in Unfamiliar Environments. *Annals of the Association of American Geographers*, 85(1):134–158.

Hirtle, S. and Hudson, J. (1995). Acquisition of Spatial Knowledge for Routes. In Gärling, T., editor, *Readings in Environmental Psychology: Urban cognition*, pages 103–113. Academic Press, New York, London.

Johnson, M. (1987). *The Body in the Mind: The Bodily Basis of Meaning, Imagination, and Reason.* University of Chicago Press, Chicago.

Kuipers, B. (1982). The "Map in the Head" Metaphor. *Environment and Behavior*, 14(2):202–220.

Lindberg E., Gaerling, T. (1983). Acquisition of Different Types of Locational Information in Cognitive Maps: Automatic or Effortful Processing. *Psychological Research*, (45):19–38.

Lynch, K. (1960). *The Image of the City.* MIT Press, Cambridge.

Neisser, U. (1976). *Cognition and Reality: Principles and Implications of Cognitive Psychology.* WH Freeman and Co. San Francisco.

Peponis, J., Conroy Dalton, R., Wineman, J., and Dalton, N. (2004). Measuring the Effects of Layout upon Visitors' Spatial Behaviors in Open Plan

Exhibition Settings. *Environment and Planning B: Planning and Design*, 31:453–473.

Portugali, J. (1996). *The Construction of Cognitive Maps*. Kluwer Academic Publishers, Dordrecht, The Netherlands.

Raubal, M., Egenhofer, M., Pfoser, D., and Tryfona, N. (1997). Structuring Space with Image Schemata: Wayfinding in Airports as a Case Study. In *Proceedings of COSIT '97, LNCS 1329*, pages 85–102. Springer-Verlag, Berlin, Germany.

Siegel, A. and White, S. (1975). The Development of Spatial Representations of Large-Scale Environments. In Reese, H., editor, *Advances in Child Development and Behavior*, Volume 10, pages 10–55. Academic Press, New York.

Thorndyke, P. and Hayes-Roth, B. (1982). Differences in Spatial Knowledge Acquired from Maps and Navigation. *Cognitive Psychology*, 14:560–589.

Chapter 7

GENETIC ALGORITHM FOR ENERGY-EFFICIENT TREES IN WIRELESS SENSOR NETWORKS

Dr. Sajid Hussain, Obidul Islam

Jodrey School of Computer Science, Acadia University, Wolfville, NS, Canada

{sajid.hussain, 079885i}@acadiau.ca

Abstract This chapter presents a genetic algorithm (GA) to generate balanced and energy-efficient data aggregation spanning trees for wireless sensor networks. In a data gathering round, a single best tree consumes lowest energy from all nodes but assigns more load to some sensors. As a result, the energy resources of heavily loaded nodes will be depleted earlier than others. Therefore, a collection of trees need to be used that balance load among nodes and consume less energy. The proposed GA takes these two issues in generating aggregation trees. The GA is simulated in an open-source simulator, J-sim. The simulation results show that proposed GA outperforms a few other data aggregation tree-based approaches in terms of extending network lifetime.

Keywords: Wireless sensor networks, Genetic algorithm, Energy efficient, Data aggregation trees

1. Introduction

Wireless sensor networks (WSNs) are commonly used in various ubiquitous and pervasive applications. Due to limited power resources, the energy-efficient communication protocols and intelligent data dissemination techniques are needed; otherwise, the energy resources will deplete drastically

This chapter is the extended work of the paper titled "Genetic Algorithm for Data Aggregation Trees in Wireless Sensor Networks", appeared in Proceedings of the Third International Conference on Intelligent Environments (IE), Ulm, Germany, September 24–25, 2007.

W. Minker et al. (eds.), *Advanced Intelligent Environments*, 139
DOI 10.1007/978-0-387-76485-6_7, © Springer Science+Business Media, LLC 2009

and the network monitoring will be severely limited (Schurgers and Srivastava, 2001; Abidi et al., 2000). The network lifetime is defined as the number of messages that can be received until all the nodes are in working condition. In a data aggregation environment, the gathered data are highly correlated and each node is capable of aggregating any incoming messages to a single message and reduce data redundancy. As a result all nodes build an aggregation tree directed and rooted at the base station. In this chapter, such a GA-based data aggregation tree is used where the sensors receive data from neighboring nodes, aggregate the incoming data packets, and forward the aggregated data to a suitable neighbor. Base station is a powerful node and it is connected to both WSN and traditional IP network. All data packets are directed toward base station, which receives aggregated data from the entire WSN. Although data aggregation can use data correlation and data compression techniques (Erramilli et al., 2004), a scenario is assumed where packets are simply concatenated, provided the aggregated size is less than the maximum packet size. The proposed technique would be suitable for a homogeneous WSN, where there is some spatial correlation in the collected data.

In this chapter, a genetic algorithm (GA) is used to create energy-efficient data aggregation trees. For a chromosome, the gene index determines a node and the gene's value identifies the parent node. The single-point crossover and mutation operators are used to create future generations. A repair function is used to avoid invalid chromosomes, which would contain cycles (loops). The chromosome fitness is determined by residual energy, transmission and receive load, and the distribution of load. The population size and the number of generations are based on the network size. An open-source simulator J-Sim, http://www.j-sim.org, is used for the simulation and performance evaluation.

The remainder of this chapter is organized as follows: Section 2 briefly describes the related work. Section 3 discusses the problem statement. Section 4 gives the details of using GA to determine data aggregation trees. Section 5 provides the simulation results. Finally, Section 6 concludes the chapter and gives directions for future work.

2. Related Work

There are several approaches to reduce the overall energy consumption. Kalpakis et al. (2002) proposes a maximum lifetime data gathering algorithm called MLDA. Given the location of each node and base station, MLDA gives the maximum lifetime of a network. MLDA works by solving a linear program to find edge capacities that flow maximum transmissions from each node to base station.

Dasgupta et al. (2003) improves MLDA by using a cluster-based heuristic algorithm, CMLDA, which works by clustering the nodes into groups of a given size. Then, each cluster's energy is set to the sum of the energy of the contained nodes. The distance between clusters is set to the maximum distance between any pair of nodes of two clusters. After the cluster formation, MLDA is applied among the clusters to build cluster trees. Then, CMLDA uses energy balancing strategy within a cluster tree to maximize network lifetime. CMLDA has a much faster execution time than MLDA; however, it is not suitable for non-dense networks.

Özgür Tan and Körpeoğlu (2003) proposes two minimum spanning tree-based data gathering and aggregation schemes to maximize the lifetime of the network, where one is the power aware version of the other. The nonpower aware version (PEDAP) extends the lifetime of the last node by minimizing the total energy consumed from the system in each data gathering round, while the power aware version (PEDAPPA) balances the energy consumption among nodes. In PEDAP, edge cost is computed as the sum of transmission and receiving energy. In PEDAPPA, however, an asymmetric communication cost is considered by dividing PEDAP edge cost with transmitter residual energy. A node with higher edge cost is included later in the tree which results in few incoming messages. Once the edge cost is established, routing information is computed using Prim's minimum spanning tree rooted at base station. The routing information is computed periodically after a fixed number of rounds (100). These algorithms assume all nodes perform in-network data aggregation and base station is aware of the location of the nodes.

In Hussain and Islam (2007), EESR is proposed that uses energy-efficient spanning tree-based multi-hop to extend network lifetime. EESR generates a transmission schedule that contains a collection of routing trees. In EESR, the lowest energy node is selected to calculate its edge cost. A node calculates its outgoing edge cost by taking the lower energy level between sender and receiver. Next, the highest edge cost link is chosen to forward data toward base station. If the selected link creates a cycle, then the next highest edge cost link is chosen. This technique avoids a node to become overloaded with too many incoming messages. The total procedure is repeated for the next lowest energy node. As a result, higher energy nodes calculate their edge costs later and receive more incoming messages, which balances the overall load among nodes. The protocol also generates a smaller transmission schedule, which reduces receiving energy. The base station may broadcast the entire schedule or an individual tree to the network.

The fast-growing applications of WSNs have made a significant contribution to the problem of finding strategies to reduce energy consumption of the resource constraint sensor nodes. The use of a GA in WSNs for efficient communication has been investigated in several research studies.

Khanna et al. (2006) propose a GA-based clustering approach to optimize multi-hop sensor networks. The genetic algorithm is used to generate the optimal number of sensor clusters with cluster-heads (CHs). The purpose of the system is to minimize power consumption of the sensors while maximizing coverage. Furthermore, the GA is used to dynamically create various components such as cluster-members, CHs, and next-cluster, which are then used to evaluate the average fitness of the system. Once clusters are formed, the GA also discovers low-cost path from cluster-head to sink.

Jin et al. (2003) use a GA to form a number of predefined clusters to reduce the total communication distance. Compared to direct transmission, this cluster formation also decreases communication distance by 80%. Their result showed that the number of cluster-heads is about 10% of the total number of nodes in the network.

Ferentinos et al. (2005) extend the GA proposed by Jin et al. (2003). The focus of their work is based on the optimization properties of genetic algorithm. In order to achieve better performance, the fitness function has been improved and GA parameters such as population size, probabilities of crossover, and mutation are tuned. This resulted in the reduction of energy consumption and improved the uniformity of measurement points.

To increase the network lifetime, Hussain et al. (2007) propose a GA-based energy-efficient hierarchical clustering technique. Their proposed GA determines the energy-efficient clusters, then the cluster-heads choose their associates for further improvement. However, the objective of the technique is to delay the last node death. It is assumed that sensors are one hop away from their cluster-heads and each cluster-head is also one hop away from the base station. These assumptions may not be suitable for a homogeneous network environment where all sensor nodes have identical energy resources.

Islam and Hussain (2006) propose a GA to determine frequencies of a collection of energy-efficient aggregation trees. The GA is used to maximize the network lifetime in terms of first node death. A set of n aggregation trees are given as the input to the GA. These aggregation trees are generated according to the increasing cost of the trees. The cost of a tree is the sum of the communication costs of all nodes. First, the best minimum spanning tree is generated from the undirected graph. Next, the second best minimum tree is computed. This process continues until n trees are generated. The value of n is adjusted according to the number of nodes in the network. Once all trees are generated, GA determines appropriate frequency for each tree so that network lifetime is maximized. As the number of nodes in the network increases, this technique fails to balance load among nodes which results in low network lifetime.

In this chapter, GA is used to create multi-hop spanning trees and it is not based on hierarchical clusters. The data aggregation trees created by the proposed GA technique are more energy efficient than some of the current data aggregation tree-based techniques. Furthermore, as our GA uses data aggregation spanning trees, it is only compared with other data aggregation spanning tree approaches such as PEDAPPA and EESR, as given in Section 5. The frequency of usage of data aggregation tree is fixed and it is not dynamically adjusted.

3. Problem Statement

In this work, we consider a sensor network application where each sensor is capable of aggregating multiple incoming data packets with its own data and forwards a single data packet of fixed length. Each node is equipped with small amount of energy and the objective is to keep all sensors working as long as possible.

DEFINITION 1 *In a data gathering round, base station receives data from all nodes. Each sensor acquires the required data samples for its environment, aggregates any incoming packets from its neighbors, and forwards the aggregated packet to its parent or base station.*

DEFINITION 2 *An aggregation tree forms a spanning tree rooted at base station. It contains routing information for all nodes in the network.*

DEFINITION 3 *The network lifetime is defined as the number of rounds until all nodes are in a working condition.*

DEFINITION 4 *Given a network of* n *sensor nodes* $s_1, s_2, ..., s_n$, *and an aggregation tree* T, *the load of a sensor node* s_i *in aggregation tree* T *is defined as the energy required to receive incoming packets and transmit the aggregated data to its parent node in a single round. The load of a sensor node* s_i *is denoted by* $Load_i$.

4. Genetic Algorithm (GA)

According to Goldberg et al. (1989), GA is commonly used in applications where search space is huge and the precise results are not very important. The advantage of a GA is that the process is completely automatic and avoids local minima. The main components of GA are crossover, mutation, and a fitness function. A chromosome represents a solution in GA. The crossover operation is used to generate a new chromosome from a set of parents while the mutation operator adds variation. The fitness function evaluates a chromosome based on predefined criteria. A better fitness value of a chromosome increases its survival chance. A population is a collection of chromosomes. A new population is obtained using standard genetic operations such as single-point crossover, mutation, and selection operator.

As a GA is relatively computation intensive, this chapter proposes executing the algorithm only at the base station. The proposed GA is used to generate balanced and energy-efficient data aggregation trees for wireless sensor networks. The following sections present the design of the proposed GA.

4.1 Gene and Chromosome

A chromosome is a collection of genes and represents a data aggregation tree for a given network. Each chromosome has fixed length size, which is determined by the number of nodes in the network. A gene index represents a node's identification number (ID) and the gene value indicates the node's parent ID.

gene index	0	1	2	3	4	5	6	7	...	99
gene value	70	30	70	10	12	18	25	10	...	80

Figure 1. Chromosome example.

Figure 1 shows a chromosome representation for a network of 100 nodes. The first node's ID is 0 (gene index) and its parent's ID is 70 (gene value). Similarly, the last node's parent ID is 80. Although the gene value is denoted as decimal, it can be represented in binary format. For instance, the chromosome given in Figure 1 can be represented as 1000100, 0011110, 1000100, 0001010, 0001100, 0010010, 0011001,0001010, $\cdots$ 1010000. However, regardless of decimal or binary representation, for all genetic operations, the entire gene value is treated as atomic. Furthermore, a chromosome is considered to be invalid if there are direct or indirect cycles (loops).

4.2 Population

A population is a collection of chromosomes. A new population is generated in two ways: steady-state GA and generational GA. In steady-state GA, a few chromosomes are replaced in each generation; whereas the generational GA replaces all the current chromosomes in each generation. A variation of generational GA is called elitism, which copies a certain number of the best chromosomes from the current generation to the new generation. A new generation is created using crossover and mutation operations.

In case of the proposed GA, a population is a collection of possible aggregation trees. For the *initial population*, the parent nodes are selected arbitrarily and the validity of chromosomes is also maintained. The size of the population remains fixed for a specific experiment and remains the same for all generations. The proposed GA uses the elitism technique to retain the best chromosome in each generation. Furthermore, the rest of the chromosomes are replaced by using crossover and mutation operations.

4.2.1 Fitness. According to Kreinovich et al. (1933), in nature, fitness of an individual is the ability to pass on its genetic material. This ability includes an individual's quality to survive. In GA, the fitness of a chromosome is evaluated using a defined function to solve a problem. A chromosome with a higher value has the better chance of survival.

4.2.2 Selection. The selection process determines which chromosomes will mate (crossover) to produce a new chromosome. A chromosome with a higher fitness value has a better chance of selection for mating. Several selection methods exist to select chromosomes for mating such as "Roulette-Wheel" selection, "Rank" selection, and "Tournament" selection. This chapter uses Tournament selection, where a pair of chromosomes are chosen randomly from the current population. Between these two chosen chromosomes, the more fit chromosome is selected with a predefined probability p; whereas the other chromosome is selected for mating with the probability (1−p), as described by Goldberg et al. (1989).

4.2.3 Crossover. Crossover is known as a binary genetic operator acting on a pair of chromosomes. Crossover recombines genetic material of two participating chromosomes. It simulates the transfer of genetic inheritance during the sexual reproductive process. The crossover result depends on the selection process made from the population.

This chapter uses a *crossover* operation where the crossover point is randomly selected and the gene values of participating parents are flipped to create a pair of child chromosomes. Figure 2 shows the crossover operation between participating parents of a network size of six nodes. The double vertical line shows the crossover point and the genes after crossover point are in bold font. The pair of children is produced by flipping the gene values of the parents after the crossover point. In this example, the last two genes are flipped.

4.2.4 Repair Function. Notice that the crossover operation between two valid chromosomes may produce invalid chromosome(s). A repair function is used to identify and prevent the inclusion of invalid chromosomes in the new generation. Figure 3 shows a pseudo-code for the repair function. For each gene in the chromosome, it checks whether the gene forms a cycle or not. If a cycle is found, a random node is selected as a potential parent. If the potential parent also forms a cycle, the process of finding a new parent continues until the end of the chromosome is reached.

Parent 1

gene index	0	1	2	3	4	5
gene value	1	2	5	4	**1**	**null**

Parent 2

gene index	0	1	2	3	4	5
gene value	1	2	5	1	**5**	**null**

Child 1

gene index	0	1	2	3	4	5
gene value	1	2	5	4	**5**	**null**

Child 2

gene index	0	1	2	3	4	5
gene value	1	2	5	1	**1**	**null**

Figure 2. Crossover example.

procedure $RepairFunction(chromosome)$
1: **for** each $gene_i$ in chromosome **do**
2: **while** $gene_i$ creates a cycle **do**
3: randomly select a new parent for $gene_i$
4: **end while**
5: **end for**

Figure 3. Chromosome repair function.

4.2.5 Mutation. The *mutation* adds variation in the next generation. In mutation, a node is randomly picked and its parent ID is reselected randomly. Similar to crossover, the mutation operation may produce an invalid chromosome, which is also fixed using the repair function.

4.3 Fitness Parameters

The fitness parameters are designed with two objectives. First, we need an energy-efficient aggregation tree so that nodes can communicate for a longer period of time. Second, the generated tree should balance load among nodes. A few fitness parameters are described in this section.

4.3.1 Ratio of Energy to Load (L). The load of a sensor node represents the communication energy required in a single round as previously described in Definition 4. A sensor node has different loads in different aggregation trees. The tree generated by the GA needs to be energy efficient. It is preferable that each node spend less amounts of energy in a communication round. For a node i, the term $\left(\frac{e_i}{Load_i}\right)$ denotes the energy consumption ratio in the tree. We want to maximize this ratio for all nodes.

4.3.2 Residual Energy of the Weakest Node (R_{min}).

An aggregation tree may assign extra burden to a specific node which may cause smaller network lifetime. The term $R_{\min} = \min(e_i - Load_i)$ denotes the energy situation of the weakest node after executing a tree. The higher value of $R_{\min}$ of a chromosome indicates that the corresponding tree has not overassigned load to any node.

4.3.3 Edge Weight (E_weight).

Edge weight fitness ensures that weak nodes avoid receiving any incoming messages. In an aggregation tree, several neighbor nodes may select the same low-energy parent node which may result in a lower lifetime value for the parent node. The E_weight fitness parameter ensures that neighbor nodes avoid selecting a low-energy node as a potential parent. It also emphasizes that a transmitting node spends a small amount of energy to transfer data to its parent.

The computation of E_weight works as follows. After a chromosome network is generated and repaired, the residual energy of each node in the chromosome is estimated for the next data gathering round using the energy model described by Heinzelman et al. (2000). Next, the nodes are sorted according to the residual energy and the edge weight is computed for each node by taking the minimum energy between the sending and the receiving node. Finally, this weight is divided by the index of the node in the sorted list.

Figure 4 shows the algorithm of the edge weight fitness calculation. The E_weight function receives a chromosome C. The residual energy of each node in the next round is computed at line1 and this information is stored in a list V. At line2, nodes are sorted in ascending order according to their energy level. For each node u in list V, its parent node p is computed at line5 and

```
procedure E_weight(C)
1: V ← computeNextRoundResidualEnergy (C)
2: V ← sort (V)
3: totalWeight ← 0
4: for each node u ∈ V do
5:    p ← getParent (u, V)
6:    weight ← min (u.energy, p.energy)
7:    totalWeight ← totalWeight + weight * 1/i
8: end for
9: return totalWeight
```

Figure 4. Edge weight fitness.

the edge weight is computed at line6. Line7 adds each node's edge weight and computes the final weight.

The E_weight fitness parameter ensures that if a weak node has a large number of children from its neighbors, then all the children of the weak node will receive low edge weight. This will cause low fitness of the chromosome network. The multiplication factor in line7 gives higher weights to weaker nodes. This edge weight follows the strategy of the EESR edge weight as given below.

The E_weight parameter can be considered as a subset of the EESR edge weight. In EESR edge weight assignment, for each transmitting node, the weight of all the edges for its neighbor is computed. For example, if a transmitting node has $(n - 1)$ neighbors, where n is the number of nodes in the network, EESR computes weight of $(n - 1)$ edges for that node and selects the highest weighted neighbor as a parent. On the other hand, for the same network, the GA edge weight parameter (E_weight) computes weight for total n edges, a single edge weight for each node in the chromosome. Moreover, E_weight is computed to evaluate a tree whereas the EESR edge weight is computed to generate an aggregation tree.

4.4 EESR Edge Weight Assignment

The aggregation tree generation algorithm starts with assigning edge weight among nodes and forms a spanning tree. To transmit a *k*-bit packet from node *i* to node *j*, the weight assignment is performed as follows:

$$w_{ij}(k) = min\{E_i - T_{ij}(k), E_j - R_j(k)\} \tag{7.1}$$

where E_i is the current energy of node *i* and $T_{ij}(k)$ is the energy required to transmit a *k*-bit packet from node *i* to node *j*. The term $R_j(k)$ denotes energy consumed to receive a *k*-bit packet for node *j*. Both $T_{ij}(k)$ and $R_j(k)$ can be computed by the energy consumption model described by Heinzelman et al. (2000). Notice that the edge weight function is asymmetric which considers the residual energy of both the sender and the receiver nodes.

4.4.1 EESR Edge Weight Example. Figure 5 shows an example of asymmetric edge weight assignment between a pair of nodes. The term E denotes initial energy of a node, while R denotes residual energy of that node after a fixed length packet has been communicated. The transmission cost between $node_1$ and $node_2$ is assumed to be symmetric and a value of 0.2 unit considered; the receiving cost for both nodes is assumed 0.1 unit. The initial energy of $node_1$ and $node_2$ is 2.3 and 2 unit, respectively. In Figure 5(a), the edge cost to send a fixed length packet from $node_1$ to $node_2$ is computed. For both nodes, the residual energy after transmission is estimated.

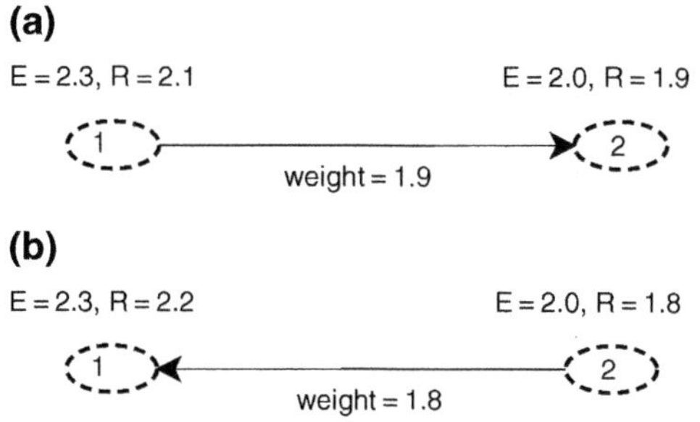

Figure 5. Example of asymmetric edge weight. (**a**) Edge cost to send fixed packet length (Tx = 0.2 and Rx = 0.1); (**b**) Edge cost to send fixed packet length (Tx = 0.2 and Rx = 0.1).

After transmission, the residual energy of $node_1$ and $node_2$ becomes 2.1 and 1.9 unit, respectively, and a minimum of them is taken as the edge weight (1.9). Similarly, if $node_2$ sends the same length packet to $node1$, the residual energy of both nodes becomes 1.8 and 2.2 unit, respectively, and the minimum (1.8) is taken as the edge weight.

4.5 Fitness Function

The above-mentioned fitness parameters are evaluated for a chromosome c_i and a fitness score f_i is calculated as follows:

$$f_i = w_1 \cdot L + w_2 \cdot E\_weight + w_3 \cdot R_{min} \tag{7.2}$$

where w_1, w_2, and w_3 are weight factors and are updated as follows:

$$w_{current} = \frac{w_{previous} + |f_{current} - f_{previous}|}{1 + e^{-f_{previous}}} \tag{7.3}$$

In Equation (7.3), $w_{current}$ and $w_{previous}$ are the current and previous weights, respectively. Similarly, $f_{current}$ and $f_{previous}$ are the fitness values of the current and previous best chromosomes. The initial value of the weight factors is adjusted using the technique described in earlier work by Hussain et al. (2007). Each chromosome is evaluated based on the given fitness criteria. The most fit ones have higher chances to survive for the next generation.

4.6 Stopping Criteria and Population Size

Determining a stopping criteria and an appropriate population size are two important factors in the design of a genetic algorithm. Two experiments are performed to determine the proposed GA's stopping criteria and population size. A network area of 100×100 m^2 with 100 sensors is considered. The base station is placed away from the network field (200, 200). In both experiments, the fitness value is taken as the decision criteria. To be confident in the experiment result, three different random networks are considered and their average is taken.

To determine a suitable population size, an experiment varying the population size for a fixed number of generations (200) is performed. The graph in Figure 6(a) is obtained by plotting the best fitness value for each experiment.

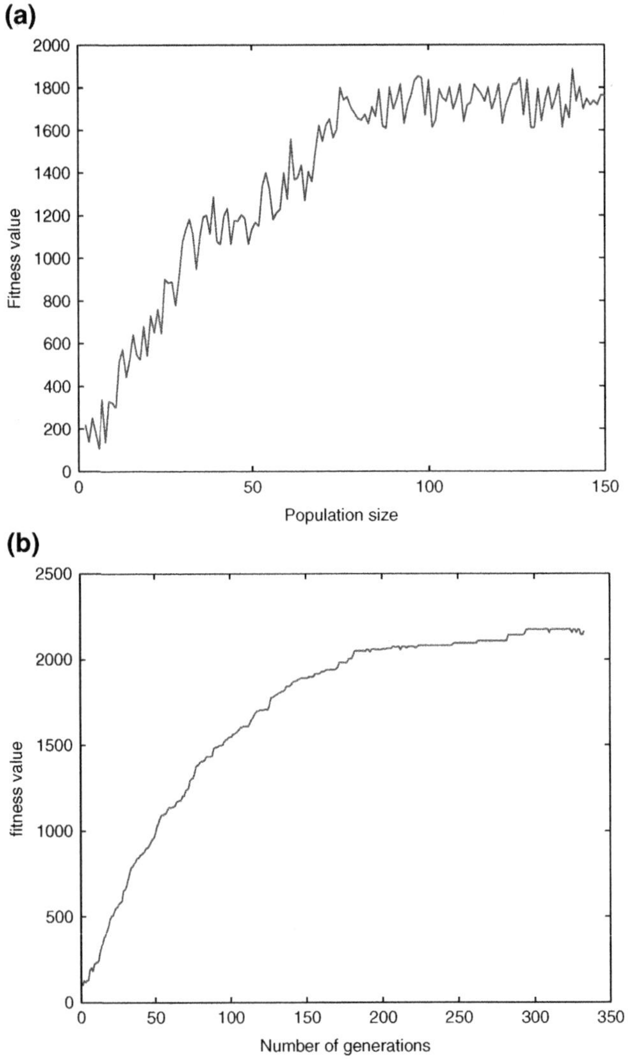

Figure 6. Impact of population size and number of generations on fitness. **(a)** Impact of population size on fitness value; **(b)** Impact of number of generations on fitness value.

The number of generations remained the same for all experiments, while the population size is fixed for a single experiment and then increased by 1 in the next trial. Moreover, the graph in Figure 6(a) shows that as the population size increases, the best fitness value gets higher. Furthermore, when the population size nearly reaches value 100, which is also the number of nodes in a chromosome, the increasing trend of fitness value diminishes. For this reason, the population size is chosen to be directly proportional to the number of nodes.

To determine the GA's stopping criteria, another experiment is performed varying the number of generations for a fixed population size (100). The graph in Figure 6(b) shows the relationship between the fitness value and the number of generations. The graph is obtained by drawing the fitness value of the best chromosome of each generation for the same experiment. As the number of generations increases, the fitness value gets higher. This is because the weak chromosomes are filtered by the fitness functions in each generation. It can be observed that the fitness value becomes stable near generation number 200, which is twice the number of nodes in the network. The stopping criteria of the proposed GA is decided by a binary function S, which is obtained as follows:

$$S(g, x) = \begin{cases} \text{true} & \text{if } |g| \geq 2 * n \text{ and } \sigma(g, x) < 0.01 \\ \text{false} & \text{otherwise} \end{cases} \tag{7.4}$$

where g is the current generation, x is a number which represents the last consecutive number of generations prior to the gth generation, and n is the number of nodes in the network. Given the values of g and x, the stopping function S returns true, if the current generation number is greater or equal to twice the number of nodes in the network, and the value of $\sigma(g, x)$ is less than 1%. The function $\sigma(g, x)$ gives the standard deviation of best fitness values of the last x consecutive generations prior to current generation. A very small value of $\sigma(g, x)$ indicates that the fitness values of the best chromosomes of the last x number of generations are almost same. This work uses a fixed value (10) for x. Recall that the graph in Figure 6(b) shows that the change in the fitness value is very small when the number of generation is $|g| > 2 * n$, where n is the number of nodes. The term $|g| \geq 2 * n$ ensures that the local maxima is not taken as the best fitted chromosome.

It can be also noted that the graph in Figure 6(b) is much smoother than that of Figure 6(a). This is because, in each generation, the best fitted chromosome is retained in the new generation using eliticism selection; thus fitness value improves in successive generations and the graph becomes smoother.

5. Simulation

5.1 Simulation Environment

We have used two types of simulators: customized simulator and J-Sim. As J-Sim provides complete protocol stack, the simulation of lower layers such as MAC and physical, the data transfer, packet collisions, latency, idle listening, and overhearing are incorporated in the simulation results. On the other hand, a customized simulator provides only single layer of simulation and can be used to verify an algorithm's correctness. In both simulators, the nodes are deployed randomly to ensure that different networks are analyzed where experiments are repeated for a given number of nodes.

5.1.1 Customized Simulator. The customized simulator, built in Java language, has only an application layer. Wireless channel is assumed to be ideal where there was no retransmission of control packets (ACK, CTS, and RTS) due to collision. Moreover, energy required to receive schedule from base station is not simulated. As there is no packet collision and retransmission, no data delay occurred in the customized simulator. For spanning tree algorithms (`EESR, GA, and PEDAPPA`), once a tree is generated, residual energy of nodes is estimated using the communication model described by Heinzelman et al. (2000) and the next tree is computed based on that energy level. For `CMLDA`, edge capacities are calculated using the same communication model, then all trees are generated by the algorithm proposed by Kalpakis et al. (2002).

5.1.2 J-Sim Simulator. J-Sim[1] is an open-source simulator based on the component-based software architecture developed entirely in Java. J-Sim provides a modeling, simulation, and emulation framework for wireless sensor networks. In J-Sim, an entity is called a component. A component may have several ports which are used to communicate with other components. The sensor network in J-Sim contains three types of nodes: (1) target nodes, (2) sensor nodes, and (3) sink nodes. Target nodes are the source nodes that generate an event in the network. A target node delivers a generated event in the sensor channel using its sensor stack. Sensor nodes have two stacks: (1) wireless protocols stack and (2) sensor stack. A sensor node receives any event from sensor channel using sensor stack, processes that information, and forwards the processed information through wireless channel using wireless protocols stack. Sink nodes have only wireless protocols stack, which receives all sensor data from wireless channel. J-Sim allows a loosely coupled environment where components communicate with each other through specific ports using predefined messages called contracts. There are two types of contracts: port contract and component contract. The port contract is specific to a port, while the component contract specifies how the component reacts when data arrives at a particular port.

The J-Sim simulation design is divided into two main components: scheduler and schedule processing. We can consider the scheduler as a base station program and schedule processing as a sensor node program. The scheduler component wraps any aggregation tree generation algorithm. In other words, the scheduler encodes routing information produced by underneath aggregation tree algorithm. This encoded information is defined as schedule. The schedule processing component decodes and executes the schedule information. Next sections describe the details of how the scheduler component is implemented in J-Sim.

5.2 Scheduler

The scheduler generates a schedule using an aggregation tree algorithm (EESR, GA, PEDAPPA) at the base station, where the location information of all nodes is available. The tasks of the scheduler is divided into two parts: schedule generation and schedule dissemination. The details of these two parts are described below.

5.2.1 Schedule Generation. An online scheduling technique is used to generate schedule for the network. The online scheduler generates an aggregation tree based on the current energy level of nodes. Recall that a schedule is a collection of aggregation trees associated with their corresponding frequencies to maximize the network lifetime. However, in the case of the online scheduler, the schedule is redefined to be an aggregation tree along with its frequency. If an aggregation tree T_i is used for m number of rounds, then at the mth round, all nodes send their energy information along with their data packets. Once the scheduler receives energy information from all nodes, the T_{i+1} tree is computed based on the current energy information received from the nodes. The scheduler estimates the communication cost for each node using the energy model given by Heinzelman et al. (2000).

Figure 7 shows the interaction of the scheduler and a simulator. The scheduler provides a schedule, which contains an aggregation tree T, along with its frequency f. Given any energy model and location of nodes, the scheduler generates a schedule which is independent of any simulator. The schedule can be generated using any algorithm (EESR, GA, or PEDAPPA) and the scheduler is independent of the WSN simulation.

5.2.2 Schedule Dissemination. Once a schedule is computed, the base station broadcasts the entire schedule in a single packet or multiple packets to the network.

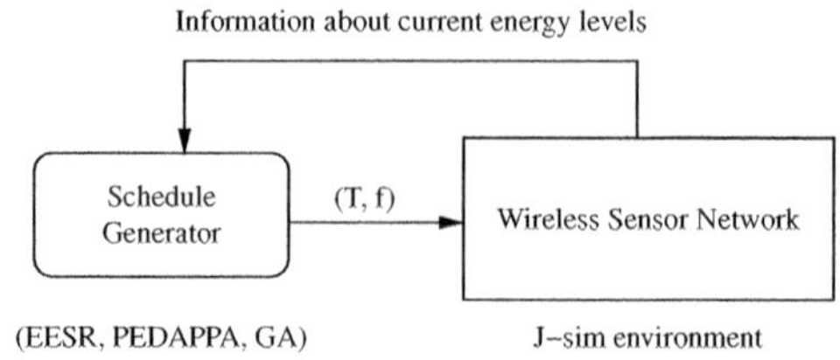

Figure 7. Interaction between scheduler and simulator.

We use a single packet format which contains routing information for all nodes of a tree. We call this single schedule packet as `SchedulePacket` which is a collection of `NodeSchedulePacket` as shown in Figure 8(b). The first byte of the `SchedulePacket` contains the frequency of the aggregation tree. The remaining part of the `SchedulePacket` contains a collection of `NodeSchedulePacket`. A `NodeSchedulePacket` contains routing information for a single node of an aggregation tree. A `NodeSchedulePacket` has three segments as shown in Figure 8(a). The first segment is 1 byte long and represents a node identification number (ID). The next segment contains parent node ID whom the node should forward its data. The last segment holds IDs of the child nodes. For a network of n number of nodes, the maximum size of the last segment of a `NodeSchedulePacket` can be $n - 2$ bytes. Experiments have shown that in a typical aggregation tree of 100 nodes, the maximum number of children of a node does not exceed over 10. In this work, we allocate fixed number of bytes (10) for children IDs in each `NodeSchedulePacket`. For a network of 10 nodes, the size of a `NodeSchedulePacket` is 12 bytes (1+1+10); therefore, the size of the `SchedulePacket` becomes ($1+12\times10$) 121 bytes. Once all nodes complete a schedule, they turn their receiver on to receive next `SchedulePacket`. The implementation details of schedule processing in J-Sim are given in Islam and Hussain (2007).

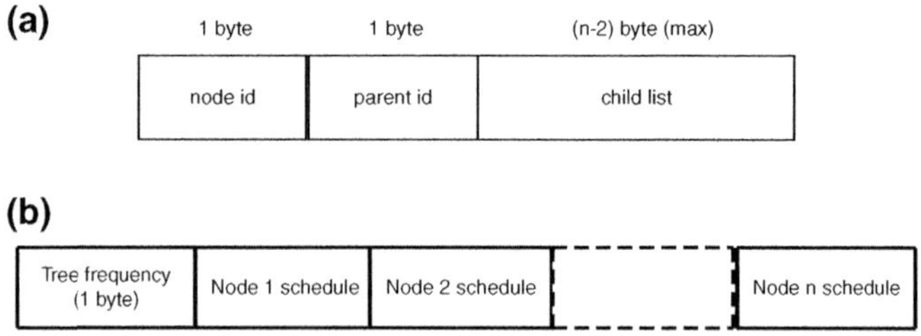

Figure 8. A single schedule packet structure. (**a**) Node schedule packet; (**b**) Schedule packet.

The schedule of data aggregation tree is created at the base station using any algorithm such as PEDAPPA, EESR, or GA. Then, base station broadcasts the schedule to the sensor network. Each sensor node extracts its own information from the schedule packet. The sensor nodes use the schedule for a given number of rounds, as specified in the schedule. At the last round of the current schedule, each node appends its residual energy level to the data packet. Then, base station uses the current energy resources to generate the next data aggregation tree.

5.3 Simulation Parameters

Table 1 provides the simulation parameters used for results and performance evaluation. Simulation parameters specific for J-Sim and GA are given in Tables 2 and 3, respectively. Table 4 provides the parameters of radio communication model used in the assessment of energy consumption in sending or receiving a message. Furthermore, this radio communication, which is originally proposed in Heinzelman et al. (2000), is commonly used in the assessment of energy consumption in WSNs and it is also used in PEDAPPA (Özgür Tan and Körpeoğlu, 2003).

Table 1. Simulation parameters.

Parameter	Remarks
Network area	$50{\times}50$ m^2 and $100{\times}100$ m^2
Number of nodes	50–100 with an interval of 10
Base station location	(a) At the center of network field and (b) outside network field at the distance of 100 m.
Data packet length	It is assumed that each sensor generates fixed length data packet of size 1000 bits.
Initial energy	Each sensor is initialized with 1 J
Radio model	First-order radio model as described in Section 5.3.
Confidence	For each experiment, sensors are placed randomly in the field and the average of five different experiments is used for performance evaluation.

Table 2. J-sim simulation parameters.

Data transfer rate	$bandwidth = 1$Mbps
Idle current	$P\_idle = 0.0002$ A
Frequency	$freq = 914$ MHz
Receiver threshold	$RXThresh = 6$ nW
Carrier sense threshold	$CSThresh = 1$ nW
Capture threshold	$CPThresh = 10$ db

Table 3. GA simulation parameters.

Population size	Population size is equal to the number of nodes
Number of generations	Population size is equal to the number of nodes
Mutation rate	0.006
Crossover rate	0.8
Tournament selection	Probability of 0.9

Table 4. Parameters of radio communication model.

Transmitter electronics	$e_{elec} = 50$ nJ/bit
Receiver electronics	$e_{elec} = 50$ nJ/bit
Transmitter amplifier	$e_{amp} = 100$ pJ/bit/m^2
Receive energy consumption	$R = e_{elec}l$, where l is the length of packet (bits)
Transmit energy	$T_{ij} = l(e_{elec} + e_{amp}d_{ij}^2)$
Initial energy of a node	1 J

5.4 Network Lifetime in Custom Simulator

This section presents the performance of EESR and GA aggregation tree algorithm in custom simulator and compares with CMLDA and PEDAPPA. To get the confidence of the results, both network fields dense and sparse network field are considered. To see the variation in network lifetime due to base station location, all experiments are performed while the base station is placed at the center and away from the network field.

5.4.1 Network Lifetime in Dense Network Field. In this experiment, the proposed techniques are investigated in a dense network field, 50 m × 50 m. We compare the network lifetime and the number of distinct routing trees generated by EESR, GA, CMLDA, and PEDAPPA protocols.

(a)

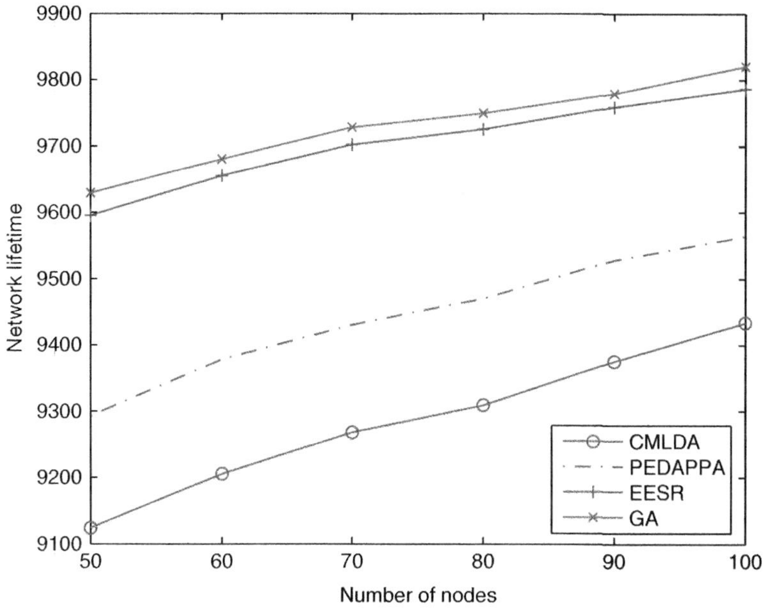

(b)

Nodes	*CMLDA*	*PEDAPPA*	*EESR*	*GA*
50	9120	92	83	85
60	9158	92	84	85
70	9236	93	85	84
80	9286	94	86	85
90	9306	95	86	86
100	9335	95	87	85

Figure 9. Simulation results using custom simulator for 50 m $\times$ 50 m network field, BS is at the center $(25, 25)$. **(a)** Variation in the network lifetime (number of transmissions) with respect to the number of nodes; **(b)** Tree frequency for CMLDA, PEDAPPA, EESR, and GA for different number of nodes.

Figures 9(a) and 10(a) show the simulation results when base station is within and outside of the network, respectively. In both cases, the proposed techniques (`EESR` and `GA`) outperform `CMLDA` and `PEDAPPA` in terms of network lifetime.

When the base station is placed at the center of the network field, `GA` gives better lifetime than `EESR`. It is noted that when the base station is at the center of the network it forms a star-like network topology, where each node tries to

(a)

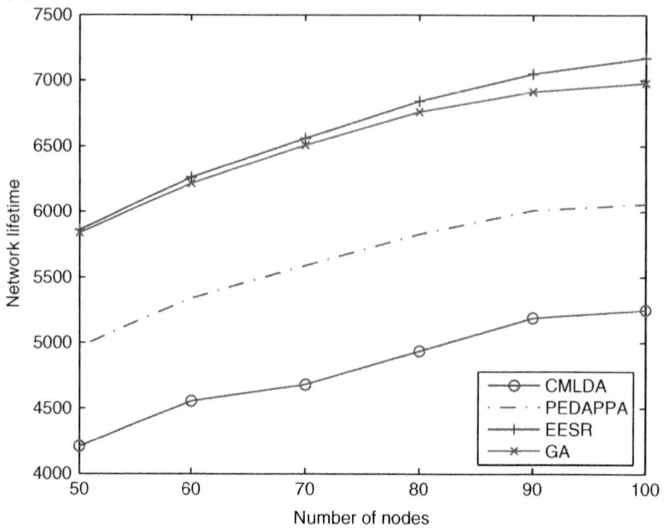

(b)

Nodes	*CMLDA*	*PEDAPPA*	*EESR*	*GA*
50	4220	51	233	229
60	4534	54	241	236
70	4637	55	249	244
80	4756	56	252	245
90	4816	57	266	257
100	5002	58	274	270

Figure 10. Simulation results using custom simulator for 50 m $\times$ 50 m network field, BS is outside $(150, 150)$. **(a)** Variation in the network lifetime (number of transmissions) with respect to the number of nodes; **(b)** Tree frequency for CMLDA, PEDAPPA, EESR, and GA for different number of nodes.

forward its data toward the center of the network. As a result, base station has more number of direct children and there are fewer incoming packets for several nodes. On the other hand, when the base station is away from the network, some nodes have to use long-range transmission to forward aggregated data to the distant base station, and the lifetime for all algorithms decreases as expected, as shown in Figures 10(a) and 9(a).

Furthermore, EESR gives better lifetime than GA when the base station is placed outside of the network as shown in Figure 10(a). When the base station is placed outside of the network, a few powerful nodes select base station as their parent, and remaining nodes use these powerful nodes to relay their data packets. Note that EESR edge cost function avoids a node receiving too many incoming packets; hence, it gives better performance than GA.

The performance of CMLDA varies with cluster size. For small cluster sizes, CMLDA acts as MLDA, which results in better lifetime at the cost of high time complexity. On the other hand, large cluster sizes in CMLDA result in traditional spanning trees that suffer from reduced network lifetime. For all experiments, we set the cluster size of 8 and 10 nodes and average results are taken.

Figures 9(b) and 10(b) show tree frequencies generated by different algorithms when the base station is at the center and outside of the network, respectively. A schedule with high tree frequency indicates that base station needs to broadcast aggregation tree more frequently. As a result, nodes spend more energy in receiving messages and the overall network lifetime decreases. Both EESR and GA minimize the number of distinct aggregation trees when the base station is inside of the network as shown in Figure 9(b). Recall that PEDAPPA uses fixed frequency (100) for each routing tree. CMLDA achieves good lifetime but uses each routing tree almost once; hence, it generates a large number of distinct trees. As the base station is moved away from the network, EESR and GA achieve better lifetime than PEDAPPA, with slight increase in the number of distinct trees, as shown in Figure 10(b).

5.4.2 Network Lifetime in Sparse Network Field. To experiment EESR and GA's network lifetime performance in a sparse network, 100 m $\times$ 100 m network field is chosen. Figure 11(a) presents proposed algorithms performance for 100 m $\times$ 100 m network when the base station is at the center of the network field. Similar to Figure 9(a), GA performs better than EESR when the base station is placed at the center of the field. Figure 11(b) shows number of distinct trees generated by different protocols when the base station is placed at the center of the field. Similar to Figure 9(b), both EESR and GA protocols archive higher lifetime and generate small number of distinct routing trees than PEDAPPA and CMLDA.

Figure 12(a) shows proposed algorithms performance when the base station is placed outside of the network. As more nodes are deployed, receiving cost becomes the leading factor in energy depletion and EESR performs better than others. Recall that EESR balances load among sensor nodes by considering receiving energy in edge cost function, thus performs best when the base station is placed outside as shown in Figure 12(a).

When the base station is moved away from the field, EESR and GA achieves better lifetime with small increase in distinct tree size as shown in Figure 12(b). Further more, energy consumption from a single tree increases as base station is moved away from the field. As a result, an aggregation tree should be used for fewer number of rounds to balance load among nodes. Recall, PEDAPPA

(a)

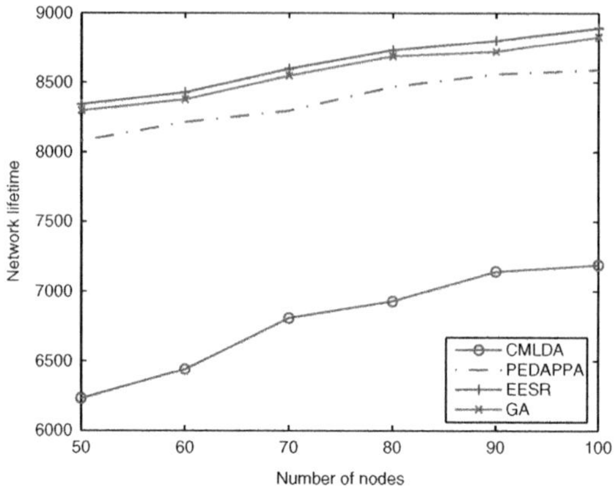

(b)

Nodes	*CMLDA*	*PEDAPPA*	*EESR*	*GA*
50	6120	81	72	78
60	6158	82	73	77
70	6236	82	74	81
80	6286	83	75	71
90	6306	83	76	79
100	6335	84	76	79

Figure 11. Simulation results using custom simulator for 150 m × 150 m network field, BS is at the center $(50, 50)$. **(a)** Lifetime in 100 m × 100 m network field, BS at $(50, 50)$; **(b)** Tree frequency in 100 m × 100 m, BS at $(50, 50)$.

uses fixed number of rounds for each routing tree. When base station is away from the network, PEDAPPA consumes large amount of energy from each tree which results in fewer distinct trees with the cost of poor network lifetime. If we compare PEDAPPA performance line in Figures 12(a) and 11(a), we can see that PEDAPPA lifetime performance is much closer to EESR and GA in Figure 11(a) than in Figure 12(a). As long-transmission range consumes more energy than short-transmission range, an aggregation tree spends more energy when the base station is placed outside of the network. As PEDAPPA uses an aggregation tree of 100 rounds, nodes drain energy at the same rate for longer period of time, which results in unbalanced energy level of nodes and shorter network lifetime. Therefore, PEDAPPA's performance line in Figure 12(a) falls further below than in Figure 11(a).

The experiment in custom simulator shows that both EESR and GA perform better than existing protocols in terms of network lifetime. In addition, the proposed technique generates fewer distinct aggregation trees. The simulation results show that GA is more suitable when the base station is at the center. On the other hand, strength of EESR can be observed when the base station is placed outside of the network.

(a)

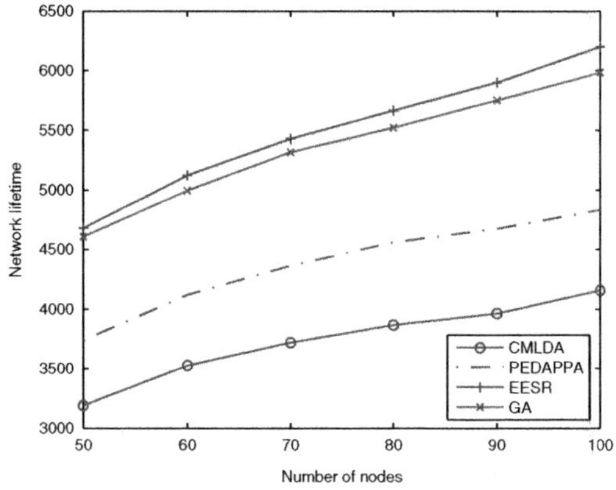

(b)

Nodes	*CMLDA*	*PEDAPPA*	*EESR*	*GA*
50	3120	37	212	202
60	3524	41	238	220
70	3713	43	256	247
80	3858	44	269	260
90	3903	45	274	265
100	3935	46	287	281

Figure 12. Lifetime using custom simulator in sparse network field. (**a**) Lifetime in 100 m × 100 m network field, BS at (200, 200); (**b**) Tree frequency in 100 m × 100 m, BS at (200, 200).

5.5 Network Lifetime in J-Sim Simulator

To see the `EESR` and `GA`'s performance in a simulator with complete protocol stack, experiments performed in custom simulator are repeated in J-Sim (1.3) simulator. The purpose of these experiments is to investigate how the proposed algorithms behave in the presence of low layer such as MAC layer. Experiments in J-Sim incorporate data retransmission, packet collisions, latency, idle listening, and overhearing which occur in real environment. At the base station, aggregation trees are constructed using the radio model (as described above). This radio model estimates energy consumption based on transmission range which is proportional to the distance between two communicating sensor nodes. But, as J-Sim simulates all real scenarios, as mentioned before, the actual energy consumption is different than the estimated consumption. As a result, finding optimal frequency of usage of each tree that fits in all situations could be a separate topic of research and can be applied independently on any algorithm. In J-Sim experiment, the frequencies of trees usage are constant(10) for all the trees, as given in `PEDAPPA`. In other words, the algorithms are evaluated based on the creation of energy-efficient data aggregation trees only. The purpose of the online scheduling is to construct aggregation trees based on current energy level of the nodes. In online scheduling, a routing algorithm repeatedly generates a new aggregation tree after specific rounds (10). Recall that CMLDA generates all possible aggregation trees each time it is executed. Therefore, in all experiments performed in J-Sim, CMLDA performance is not simulated as it is not a suitable choice for online scheduling.

5.5.1 Network Lifetime in Dense Network Field. Figure 13(a) and (b) shows EESR and GA's performance against PEDAPPA for 50 m $\times$ 50 m network field when the base station is at the center and away from the network field.

In Figure 13(a), all three protocols give almost similar performance in contrast to lifetime performance in custom simulator as shown in Figure 9(a). Although GA performs slightly better for small number of nodes (50, 60, 70), but as the network gets denser, all three protocols give similar network lifetime. It is noted that transmission cost is proportional to the distance of the communicating nodes (Heinzelman et al., 2000). When the base station is at the center, the transmission cost of nodes decreases. Moreover, in J-Sim, as more nodes are deployed, packet collusion, data retransmission, and overhearing increases. The cost incurred due to this scenario dominates over the communication cost of the nodes. Recall that the customized simulator does not incorporate the issues of MAC layer and packet collisions. As a result, the individual performance of the algorithms is not quite obvious in J-Sim when the base station is placed at the center and more nodes are deployed.

Figure 13(b) shows protocols performance in dense network when the base station is placed away from the network field. For each size of node deployment, EESR and GA perform better than PEDAPPA in terms of network lifetime. Moreover, EESR performance continues to prevail over GA performance similar to Figure 10(a). Note that, in this experiment, individual performance of the algorithms is more visible than Figure 13(a). When the base station is placed away from the network, some sensor nodes use longer transmission range to forward data to base station. Moreover, packets generated from sensors travel more hops to reach base station than experiments shown in Figure 13(a). As a result, communication cost prevails over retransmission and over-hearing cost, and the performances of algorithms are more prominent.

5.5.2 Network Lifetime in Sparse Network Field. Figure 14(a) and (b) shows proposed algorithms performance in sparse network when the base station is placed at the center and away from the network, respectively. The experiment setup is identical to customized simulator as presented in Figures 11(a) and 12(a). In all cases, EESR and GA perform better than existing

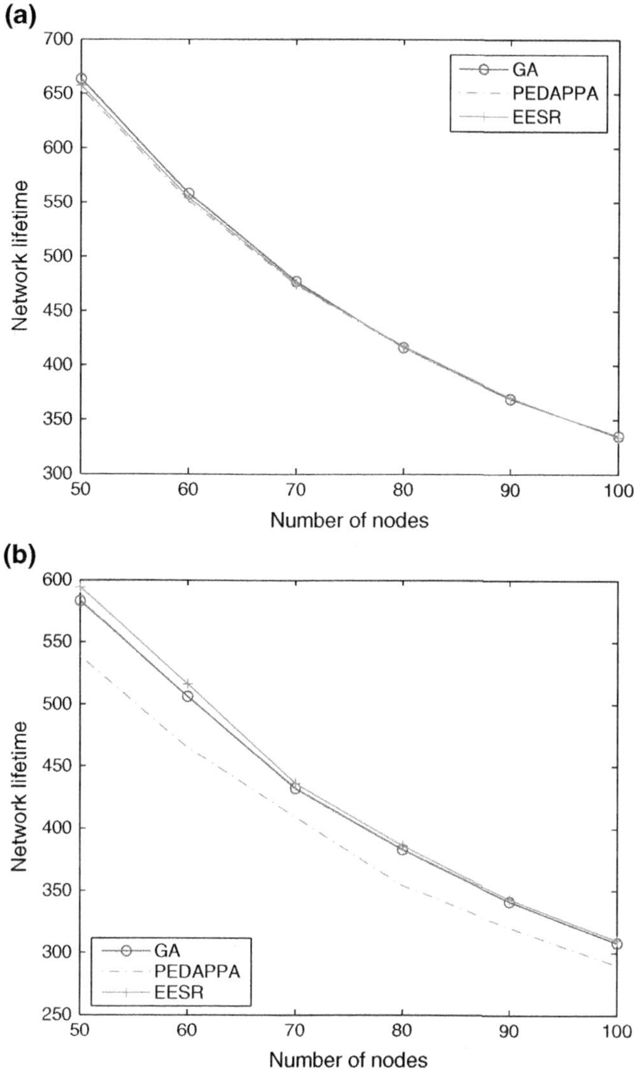

Figure 13. Lifetime using J-Sim in sparse network field. (**a**) Lifetime in 50 m × 50 m network field, BS at $(25, 25)$; (**b**) Lifetime in 50 m × 50 m network field, BS at $(150, 150)$.

aggregation tree algorithm. The relative performance of the algorithms is similar to the results obtained from customized simulator. However, one significant difference can be observed. In customized simulator, the lifetime of all algorithms increases as more nodes are deployed as shown in Figures 9 and 12. This result is expected as transmission distance becomes smaller in denser network.

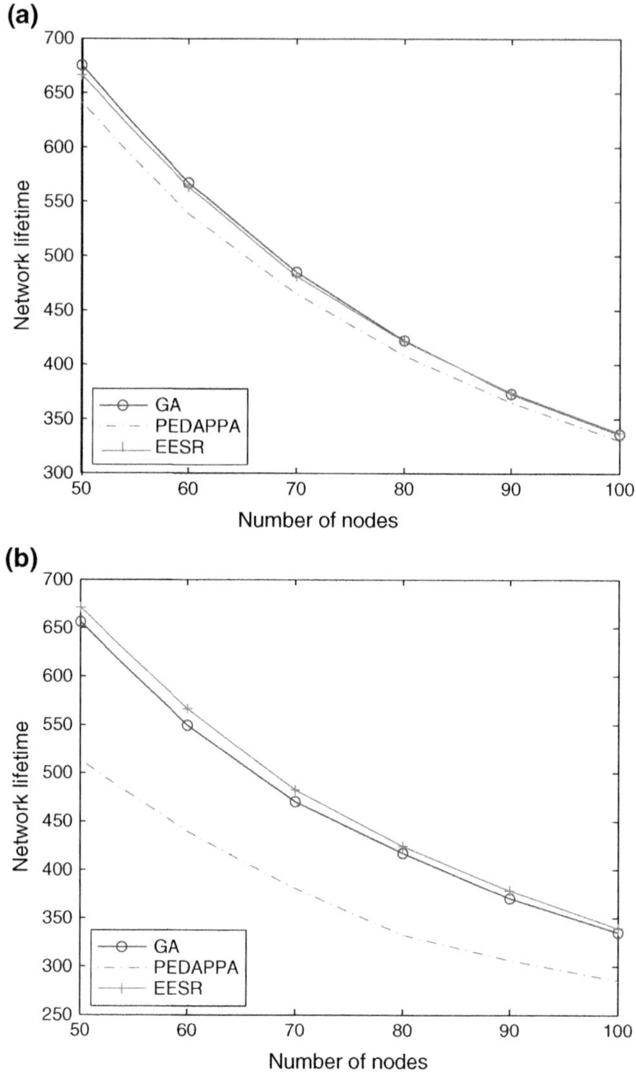

Figure 14. Lifetime using J-Sim in 100 m $\times$ 100 m field. **(a)** Lifetime in 100 m $\times$ 100 m network field, BS at $(50, 50)$; **(b)** Lifetime in 100 m $\times$ 100 m network field, BS at $(200, 200)$.

The customized simulator does not incorporate the issues of MAC layer and packet collisions; hence energy consumption due to retransmission and overhearing does not take place. Further more, the network lifetime estimated by the customized simulator is similar to Özgür Tan and Körpeoğlu (2003) for identical simulation parameters. In contrast, as J-Sim uses IEEE 802.11 MAC and remaining protocol stack, we get the effect of data collisions, retransmissions, and idle listening. Therefore, lifetime in J-Sim drops as more nodes are deployed as presented in Figures 13 and 14.

5.6 Network Lifetime in Heterogeneous Energy Level

In Figure 15, the experiment of Figure 12(a) is repeated using different initial energy for PEDAPPA, EESR, and GA algorithm. The purpose of this

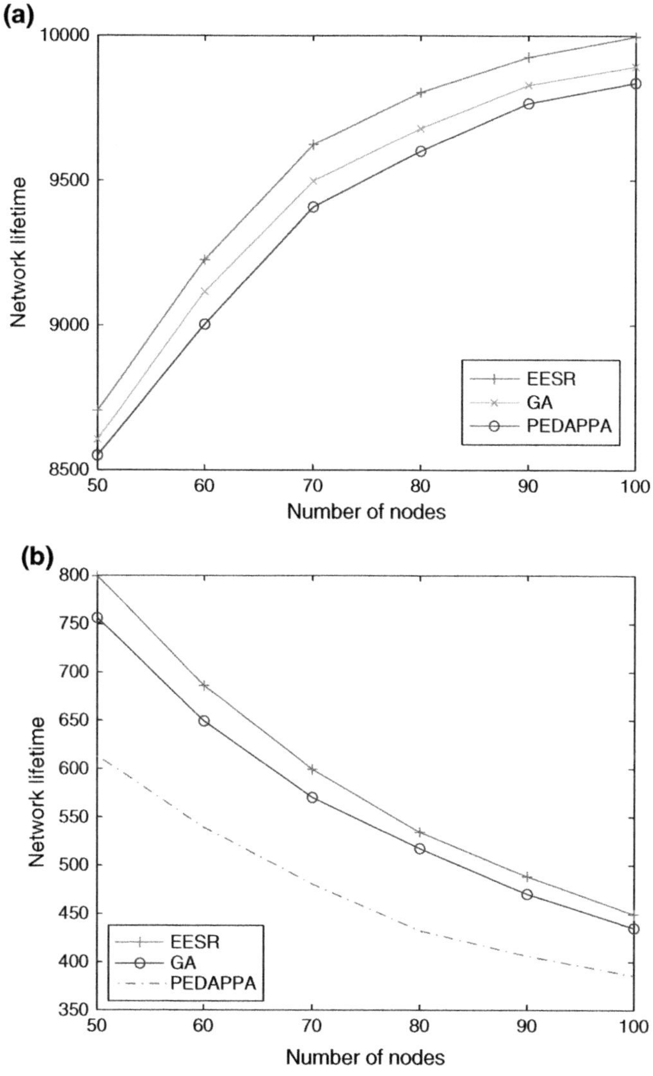

Figure 15. Lifetime performance using different energy level, 100 m×100 m field, BS at (200, 200). **(a)** Lifetime using custom simulator; **(b)** Lifetime using J-Simsimulator.

experiment is to see how the proposed algorithms behave in heterogeneous energy levels of the nodes. In this experiment, 25% of the sensor nodes are randomly selected and equipped with higher initial energy (10 J). Figure 15(a) and (b) shows performance result in custom and J-Sim simulator, respectively. In all cases, EESR, and GA outperforms PEDAPPA in network lifetime. For identical parameters, the result obtained from this experiment is similar to the experiment when all nodes have equal initial energy as shown before in Figures 12(a) and 14(b). However, the overall performance improves for all algorithms. As few nodes are more powerful than others in terms of initial energy, they take more load than rest of the nodes. Each of these powerful nodes acts as a proxy of base station or sink node which finally forwards data to the base station. Therfore, all algorithms give slightly better lifetime than the experiment performed with homogeneous energy level. Further more, EESR maintains to retain its performance over GA when the base station is away from the network field. Recall that EESR adds high-power sensors later in the aggregation tree. This causes the high-power sensor nodes to receive more incoming packets than weaker nodes and assigns more loads to them. This experiment gives the indication that EESR balances load well among sensors and can be used in heterogeneous energy level when few sensors are equipped with higher energy or new sensors are replaced in the network.

5.7 Energy Consumption Analysis

This section analyzes residual energy after the first node death. First, the effect of base station's position on nodes' residual energy is investigated. Next, we perform some statistical analysis of residual energy to see the end effect of load-balancing and energy-efficient routing protocols.

5.7.1 Impact of Base Station's Location. Figure 16 shows energy level of sensor nodes based on their position in network field after the first node death. Nodes' locations are presented in x–y plane, while z axis represents corresponding node's energy level. A network field of 100 m$\times$100 m with 50 nodes is considered. Base station is placed outside of the field at 200 m$\times$200 m. A circle indicates the presence of a low-energy node. Low-energy nodes are those nodes which caused data gathering to be terminated, or these nodes are about to die as their energy level is very low. Figure 16 and (b) shows that low-energy nodes in EESR and GA are randomly distributed in the network field. However, in Figure 16(c), which shows the results of PEDAPPA, all the low-energy nodes are near to the base station. Recall that PEDAPPA only considers the ratio of communication cost and transmitting node's energy in edge cost computation. As a result, nodes closer to base station are included in PEDAPPA tree at the very beginning. We call these nodes as gateway nodes

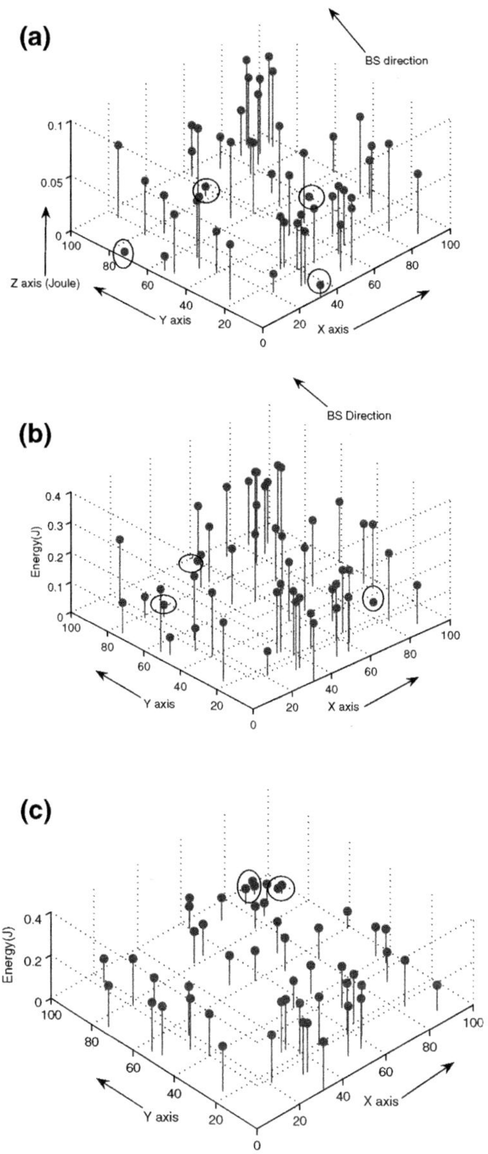

Figure 16. Energy level after the first node death, 100 m×100 m field, BS at $(200, 200)$. (**a**) EESR result; (**b**) GA result; (**c**) PEDAPPA result.

as they directly forward data to the base station. These gateway nodes are used as relay by their neighbors to forward data to the base station. As a result, all gateway nodes receive many incoming packets and consume more energy than others. Therefore, an area of low-energy nodes near the base station is observed as shown in Figure 16(c). On the other hand, EESR and GA consider both receiving and transmitting cost in edge cost function in order to balance the load among nodes. Therefore, we do not observe any specific area of low-energy nodes in Figure 16(a) and (b).

5.7.2 Uniform Consumption of Energy. A lifetime-aware routing protocol avoids uneven energy distribution among nodes. As a result, uniform residual energy level of all nodes after the first node death is expected. For uniform distribution of load, the variation of residual energy after the first node death should be small. The standard deviation of residual energy (SDR) is a good way to measure energy dispersion from the mean. A small value of SDR means that the load among nodes was uniformly distributed. The graphs in Figure 17(a) and (b) show the standard deviation of the residual energy (SDR) after the first node death for 100 m $\times$100 m field when the base station is at the center and outside of the network, respectively. In Figure 17(a), the SDR value of `GA` is smaller than others. On the other hand, in Figure 17(b), the SDR value of `EESR` is smaller than `GA` and `PEDAPPA`'s SDR values. This indicates that for the same network, `GA` performs better

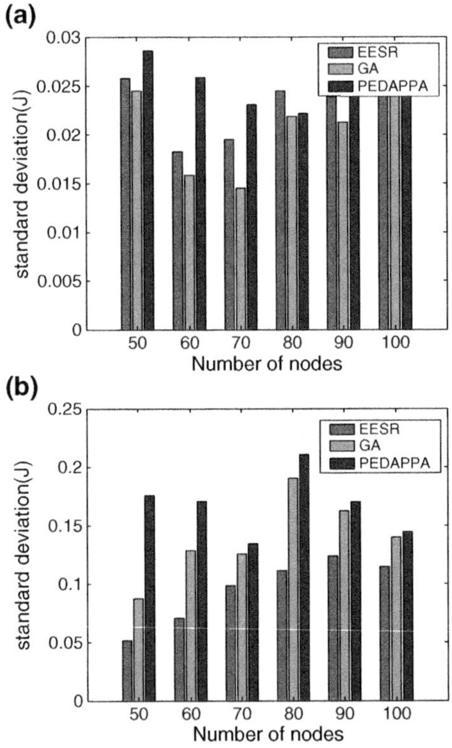

Figure 17. Standard deviation of residual energy (SDR), 100 m $\times$100 m field. **(a)** 100 m $\times$100 m field, BS at $(50, 50)$; **(b)** 100 m $\times$100 m field, BS at $(200, 200)$.

in load balancing when the base station is at the center; whereas EESR gives better performance when the base station is away from the network.

The graphs in Figure 18 show the comparison of average residual energy left in nodes after the first node death. When the base station is at the center, average residual energy (ARE) of nodes produced by GA algorithm is low as shown in Figure 18(a). In contrast, Figure 18(b) shows that executing EESR on the same network produces low ARE when the base station is away from the network. A low value of ARE means that all nodes are properly utilized by the corresponding routing algorithm.

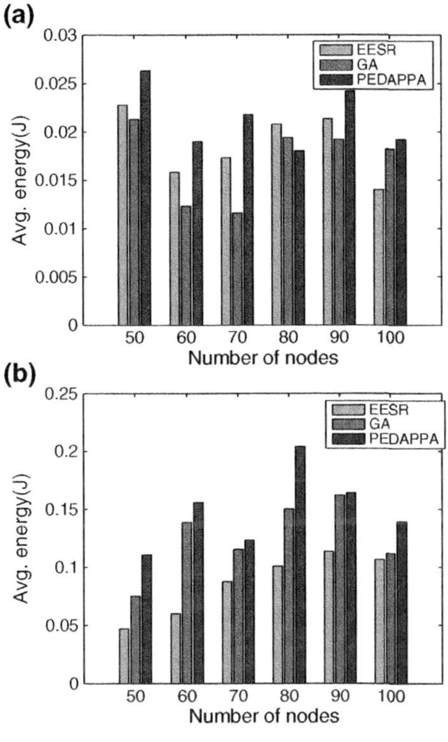

Figure 18. Average of residual energy (ARE), 100 m×100 m field. (**a**) 100 m×100 m field, BS at (50, 50); (**b**) 100 m×100 m field, BS at (200, 200).

This analysis result complements the network lifetime performance of the proposed algorithms as previously shown in Section 5.5.2. A small value of both SDR and ARE is expected to maximize network lifetime, as it indicates that the corresponding routing algorithm balances load among nodes and utilizes them as well. The smaller value of SDR and ARE of GA in Figures 17(a) and 18(a) matches the lifetime performance of GA algorithm when base station is placed at the center of the network, as shown in Figure 14(a). Similarly, as

EESR maximizes network lifetime when the base station is placed outside, the corresponding SDR and ARE values are also small as shown in Figures 14, 17(a) and 18(a), respectively.

5.8 Round Completion Time

Figure 19 shows variation in round completion time for a 100 m$\times$100 m network area. We describe *round delay* as the time to complete a round. This delay includes the sampling interval of 1 s. Notice that a round is completed when base station receives data from all sensor nodes. The graph in Figure 19(a) shows that the round delay is similar for all the algorithms when the base station is at the center of the network field. Note that placing base station at the center forms a star network topology where all packets travel a fewer number of hops to reach the base station. As a result, the average aggregation tree height is similar for all algorithms. However, Figure 19(b) shows that the round delay incurred by EESR is slightly higher than GA and PEDAPPA when the base station is placed away from the network. This result is expected because EESR performs better in load balancing when the base station is outside of the network. To balance load among nodes, EESR allows more incoming messages to higher energy nodes, while discouraging any incoming message for low-energy node. As a result, depth of the routing tree increases. It is noted that as the tree depth increases, non-leaf nodes which are close to base station have to wait more than leaf nodes, hence increases the round delay. When the network becomes dense, the depth of the routing tree increases along with collisions and retransmissions, which results in all protocols giving higher latency.

The simulation results indicate that network lifetime can be maximized by the proposed aggregation tree algorithms. Moreover, results show that GA-based aggregation tree performs better when the base station is positioned at the center of the network. GA is computation intensive as compared to EESR and PEDAPPA. However, as GA is performed at base station, it will not be an issue for a typical sensor network deployment. On the other hand, if base station is not powered by regular power source and it is like a stargate node, the proposed GA may not be a good choice. The strength of EESR can be observed when the base station is placed outside of the network. As mentioned before, the proposed protocols consider both sending and receiving nodes' residual energy in choosing routing path. PEDAPPA only considers transmitter residual energy in edge cost thus ends up assigning higher load to the receiver node and reduces overall performance. The performance of CMLDA varies with the number of clusters. Moreover, it generates a large number of distinct trees which add extra overhead in receiving schedules from base station. The simulation results also show that as number of nodes increases, the network lifetime

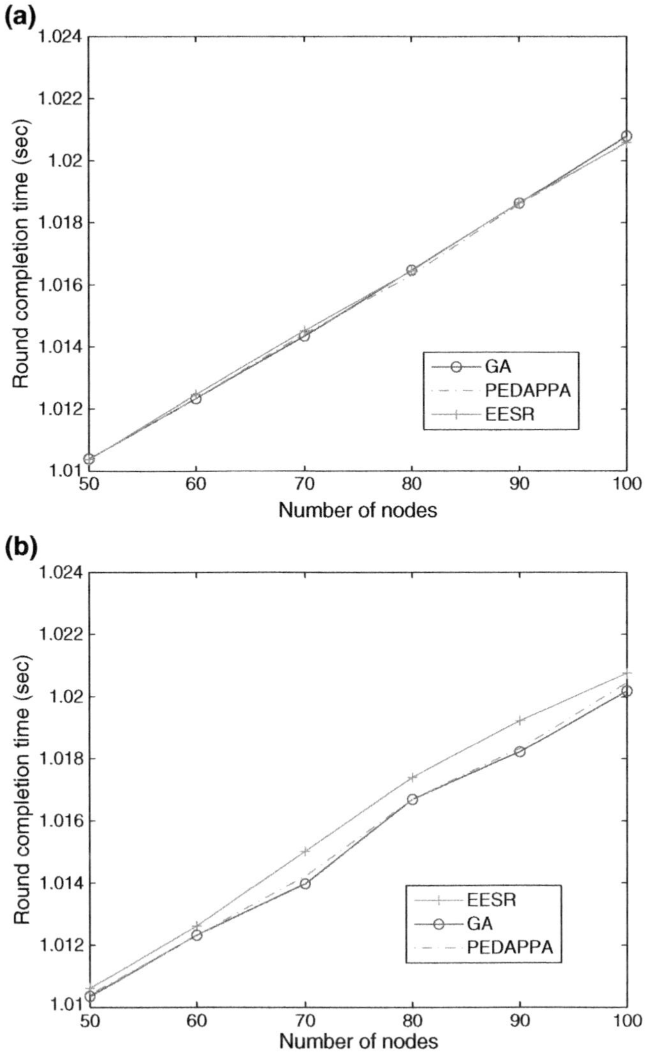

Figure 19. Time delay to complete a round. **(a)** 100 m × 100 m network field, BS at $(50, 50)$; **(b)** 100 m × 100 m network tree frequency, BS at $(200, 200)$.

varies between the customized simulator and the J-Sim simulator. This change is expected as customized simulator does not consider MAC issues. However, the relative performance of the algorithms is consistent across both simulators.

6. Conclusion

In this chapter, a genetic algorithm (GA) is used to create energy-efficient data aggregation trees. For a chromosome, the gene index determines a node and the gene's value identifies the parent node. A single-point crossover is selected to create future generations. After each crossover operation, a chromosome is checked to see if it contains cycles (loops). If cycles exist, a repair function is used to make a valid chromosome. The chromosome fitness is determined by residual energy, transmission and receive load, and the distribution of load. The population size and the number of generations remain the same for all generations. The proposed GA-based data aggregation trees extend the network lifetime as compared to EESR and PEDAPPA. Moreover, the results show that GA performs better when the number of nodes in the network is small. However, the fitness function and the remaining GA parameters can be improved or tuned to increase the network lifetime.

In future work, we would like to investigate adaptive tree frequency techniques and adaptive crossover operator to improve the performance of the algorithm. The GA-based approach could be improved by incorporating an adaptive crossover. Moreover, fitness parameters could be tuned or added to improve its performance, particularly when the base station is placed outside. This work can also be extended to maximize the network lifetime for heterogeneous network where sensed data are not highly correlated and aggregation is not possible. An implementation in TinyOS[2] would be useful to validate the proposed techniques.

Notes

1. http://www.j-sim.org/
2. http://www.tinyos.net/

References

Abidi, A. A., Pottie, G. J., and Kaiser, W. J. (2000). Power-conscious design of wireless circuits and systems. *IEEE Transactions on Mobile Computing*, 88(10):1528–45.

Dasgupta, K. Kalpakis, K., and Namjoshi, P. (2003). An efficient clustering-based heuristic for data gathering and aggregation in sensor networks. *In IEEE Wireless Communications and Networking Conference.*

Özgür Tan, H. and Körpeoğlu İ. (2003). Power efficient data gathering and aggregation in wireless sensor networks. *SIGMOD Rec.*, 32(4):66–71.

Erramilli, V., Matta, I., and Bestavros, A. (2004). On the interaction between data aggregation and topology control in wireless sensor networks. In *Proceedings of SECON*, pages 557–565.

Ferentinos, K. P., Tsiligiridis, T. A., and Arvanitis, K. G. (2005). Energy optimization of wirless sensor networks for environmental measurements.

In *Proceedings of the International Conference on Computational Intelligence for Measurement Systems and Applications (CIMSA)*.

Goldberg, D., Karp, B., Ke, Y., Nath, S., and Seshan, S. (1989). *Genetic algorithms in search, optimization, and machine learning*. Addison-Wesley.

Heinzelman, W. R., Chandrakasan, A., and Balakrishnan, H. (2000). Energy-efficient communication protocol for wireless microsensor networks. In *Proceedings of the Hawaii International Conference on System Sciences*.

Hussain, S. and Islam, O. (2007). An energy efficient spanning tree based multi-hop routing in wireless sensor networks. In *Proceedings of IEEE Wireless Communications and Networking Conference (WCNC)*.

Hussain, S. Islam, O., and Matin, A. W. (2007). Genetic algorithm for energy efficient clusters in wireless sensor networks. In *Proceedings of the 4th International Conference on Information Technology: New Generations (ITNG)*. IEEE Computer Society.

Islam, O. and Hussain, S. (2006). An intelligent multi-hop routing for wireless sensor networks. In *Workshop Proceedings of the IEEE/WIC/ACM International Conference on Intelligent Agent Technology, (IAT)*. IEEE Computer Society.

Islam, O. and Hussain, S. (2007). Effect of layers on simulation of wireless sensor networks. In *Proceedings of the 3rd International Conference on Wireless and Mobile Communications (ICWMC)*. IEEE Computer Society.

Jin, S., Zhou, M., and Wu, A. S. (2003). Sensor network optimization using a genetic algorithm. In *Proceedings of the 7th World Multiconference on Systemics, Cybernetics and Informatics*.

Kalpakis, K., Dasgupta, K., and Namjoshi, P. (2002). Maximum lifetime data gathering and aggregation in wireless sensor networks. *In IEEE International Conference on Networking*, pages 685–696.

Khanna, R. Liu, H., and Chen, H.-H. (2006). Self-organisation of sensor networks using genetic algorithms. *International Journal of Sensor Networks (IJSNET)*, 1(3/4).

Kreinovich, V., Quintana, C., and Fuentes, O. (1993). Genetic algorithms: what fitness scaling is optimal. *Cybernetics and Systems: an International Journal*, 24:9–26.

Schurgers, C. and Srivastava, M. B. (2001). Energy efficient routing in wireless sensor networks. *MILCOM*, pages 357–361.

Chapter 8

ENHANCING ANOMALY DETECTION USING TEMPORAL PATTERN DISCOVERY

Vikramaditya R. Jakkula, Aaron S. Crandall, Diane J. Cook
Washington State University, Pullman, Washington, USA
{vjakkula, acrandal, cook}@eecs.wsu.edu

Abstract Technological enhancements aid development and research in smart homes and intelligent environments. The temporal nature of data collected in a smart environment provides us with a better understanding of patterns that occur over time. Predicting events and detecting anomalies in such data sets is a complex and challenging task. To solve this problem, we suggest a solution using temporal relations. Our temporal pattern discovery algorithm, based on Allen's temporal relations, has helped discover interesting patterns and relations from smart home data sets. We hypothesize that machine learning algorithms can be designed to automatically learn models of resident behavior in a smart home and, when these are incorporated with temporal information, the results can be used to detect anomalies. We describe a method of discovering temporal relations in smart home data sets and applying them to perform anomaly detection on the frequently occurring events by incorporating temporal relation information shared by the activity. We validate our hypothesis using empirical studies based on the data collected from real resident and virtual resident (synthetic) data.

Keywords: Temporal relationships; Smart environments.

1. Introduction

The problems of representing, discovering, and using temporal knowledge arise in a wide range of disciplines, including computer science, philosophy, psychology, and linguistics. Temporal rule mining and pattern discovery applied to time series data has attracted considerable interest over the last few years. We consider the problem of learning temporal relations between event time intervals in smart environment data, which includes physical activities

W. Minker et al. (eds.), *Advanced Intelligent Environments*,
DOI 10.1007/978-0-387-76485-6_8,

(such as taking pills while at home) and instrumental activities (such as turning on lamps and electronic devices). These learned temporal relations can be used to detect anomalies. The purpose of this chapter is to identify interesting temporal patterns in order to detect whether the event which occurred is an anomaly. A simple sensor can produce an enormous amount of temporal information, which is difficult to analyze without temporal data mining techniques that are developed for this purpose.

Our vision is to keep older adults functioning independently in their own homes longer. The number of Americans who live with cognitive or physical impairments is rising significantly due to the aging of the population and better medical care. By 2040, an estimated 23% of the US population will be 65+ (Lanspery et al., 1997). Many of these elder adults live in rural areas with limited access to health care. While 90% of Americans over 60 want to live out their lives in familiar surroundings (Gross, 2007), today those who need special care must often leave home to meet medical needs. Providing this care at home will become a requirement because 40% of elder adults cannot afford to live in assisted care facilities and because hospitals and nursing homes do not have the capacity to handle the coming "age wave" of a larger, sicker population (Wang, 2006).

Data collected in smart environments has a natural temporal component to it, and reasoning about such timing information is essential for performing tasks such as anomaly detection. Usually, these events can be characterized temporally and represented by time intervals. These temporal units can also be represented using their start time and end time which lead to form a time interval, for instance when the cooker is turned on it can be referred to as the start time of the cooker and when the cooker is turned off it can be referred to as the end time of the cooker. The ability to provide and represent temporal information at different levels of granularity is an important research sub-field in computer science which especially deals with large timestamp data sets. The representation and reasoning about temporal knowledge is very essential for smart home applications. Individuals with disabilities, elder adults, and chronically ill individuals can take advantage of applications that use temporal knowledge. In particular, we can model activities of these individuals, use this information to distinguish normal activities from abnormal activities, and help make critical decisions to ensure their safety.

The objective of this research is to identify temporal relations among daily activities in a smart home to enhance prediction and decision making with these discovered relations, and detect anomalies. We hypothesize that machine learning algorithms can be designed to automatically learn models of resident behavior in a smart home and, when these are incorporated with temporal information, the results can be used to detect anomalies. We discuss Allen's notion of temporal relationships and describe how we can discover frequently occurring temporal relationships in smart home data. We then use the discovered temporal relations to perform anomaly detection. We validate our hypothesis using empirical studies based on the data collected from real resident and synthetic data.

2. Temporal Reasoning

Activities in a smart home include resident activities as well as interactions with the environment. These may include walking, sitting on a couch, turning on a lamp, using the coffee maker, and so forth. Instrumental activities are those which have some interaction with an instrument which is present and used in the environment. We see that these activities are not instantaneous, but have distinct start and end times. We also see that there are well-defined relationships between the time intervals for different activities. These temporal relations can be represented using Allen's temporal relations and can be used for knowledge and pattern discovery in day-to-day activities. These discoveries can be used for developing systems which detect anomalies and aid caregivers in taking preventive measures.

Allen (Allen and Ferguson, 1994) listed 13 relations (visualized in Figure 1) comprising a temporal logic: before, after, meets, met-by, overlaps, overlapped-by, starts, started-by, finishes, finished-by, during, contains, and equals. These temporal relations play a major role in identifying time-sensitive activities which occur in a smart home. Consider, for example, a case where the resident turns the television on before sitting on the couch. We notice that these two activities, turning on the TV and sitting on the couch, are frequently related in time according to the "before" temporal relation. Modeling temporal events in smart homes is an important problem and offers benefits to residents of smart homes. Temporal constraints can be useful when reasoning about activities; if a temporal constraint is not satisfied then a potential "anomalous" or "critical" situation may have occurred.

Temporal mining is a relatively new area of research in computer science and has become more popular in the last decade due to the increased ability of computers to store and process large data sets of complex data. Temporal reasoning and data mining have been investigated in the context of classical and temporal logics and have been applied to real-time artificial intelligence systems.

Morchen argued that Allen's temporal patterns are not robust and small differences in boundaries lead to different patterns for similar situations (Morchen, 2006). Morchen presents a time series knowledge representation (TSKR), which expresses the temporal concepts of coincidence and partial order. He states that Allen's temporal relations are ambiguous in nature, making them not scalable and not robust. Morchen handles this problem of

Temporal Relations	Pictorial Representation	Interval constraints
X Before Y		StartTime(X)<StartTime(Y); EndTime(X)<StartTime(Y)
X After Y		StartTime(X)>StartTime(Y); EndTime(Y)<StartTime(X)
X During Y		StartTime(X)>StartTime(Y); EndTime(X)<EndTime(Y)
X Contains Y		StartTime(X)<StartTime(Y); EndTime(X)>EndTime(Y)
X Overlaps Y		StartTime(X)<StartTime(Y); StartTime(Y)<EndTime(X); EndTime(X)<EndTime(Y)
X Overlapped-By Y		StartTime(Y)<StartTime(X); StartTime(X)<EndTime(Y); EndTime(Y)<EndTime(X)
X Meets Y		StartTime(Y) = EndTime(X)
X Met-by Y		StartTime(X)= EndTime(Y)
X Starts Y		StartTime(X)=StartTime(Y); EndTime(X)≠EndTime(Y)
X started-by Y		StartTime(Y)=StartTime(X); EndTime(X)≠EndTime(Y)
X Finishes Y		StartTime(X)≠StartTime(Y); EndTime(X) =EndTime(Y)
X Finished-by Y		StartTime(X)≠StartTime(Y); EndTime(X)=EndTime(Y)
X Equals Y		StartTime(X)=StartTime(Y); EndTime(X)=EndTime(Y)

Figure 1. Thirteen temporal relationships comprising Allen's temporal logic.

ambiguity by applying constraints to define the temporal relations. Although this method appears feasible, it does not suit our smart home application due to the granularity of the time intervals in smart home data sets. In smart environments, some events are instantaneous while others span a long time period. Morchen applies TSKR to muscle reflection motion and other applications where time intervals are consistently similar in length. Because the TSKR approach also does not eliminate noise and is computationally expensive, the approach is not well suited to the large and complex sensor data that is created by smart environments.

In artificial intelligence, the event calculus is a frequently used approach for representing and reasoning about events and their effects. Gottfried et al. (2006) also argue that space and time play essential roles in everyday lives and introduce time and space calculi to reason about these dimensions. They discuss several AI techniques for dealing with temporal and spatial knowledge in smart homes, mainly focusing on qualitative approaches to spatiotemporal reasoning.

Ryabov and Puuronen (2001) in their work on probabilistic reasoning about uncertain relations between temporal points represent the uncertain relation between two points by an uncertainty vector with three probabilities of basic relations ("$<$", "$+$", and "$>$"). They also incorporate inversion, composition, addition, and negation operations into their reasoning mechanism. This model would not be suitable for a smart home scenario as it would not delve into finer granularities to analyze instantaneous events. The work of Worboys and Duckham (2002) involves spatiotemporal-based probability models, the implementation of which is currently identified as future work. Dekhtyar et al.'s research on probabilistic temporal databases (Dekhtyar et al., 2001) provides a framework which is an extension of a relational algebra that integrates both probabilities and time. This work, like ours, builds on Allen's temporal logic.

3. The MavHome Smart Home

Our anomaly detection algorithm is designed as part of the MavHome smart home project (Youngblood and Cook, 2007; Youngblood et al., 2005). We view a smart environment as an intelligent agent, which determines the state of the environment using sensors and acts upon the environment using powerline controllers. All of the MavHome components are implemented and have been tested in two physical environments, the MavLab workplace environment and an on-campus apartment. Powerline control automates all lights and appliances, as well as HVAC, fans, and miniblinds. Perception of light, humidity, temperature, smoke, gas, motion, and switch settings is performed through a sensor network developed in-house. Inhabitant localization is performed using passive infrared sensors yielding a detection rate of 95% accuracy.

The MavHome architecture shown in Figure 2 consists of cooperating layers. Perception is a bottom-up process. Sensors monitor the environment using physical components (e.g., sensors) and make information available through the interface layers. The database stores this information while other information components process the raw information into more useful knowledge

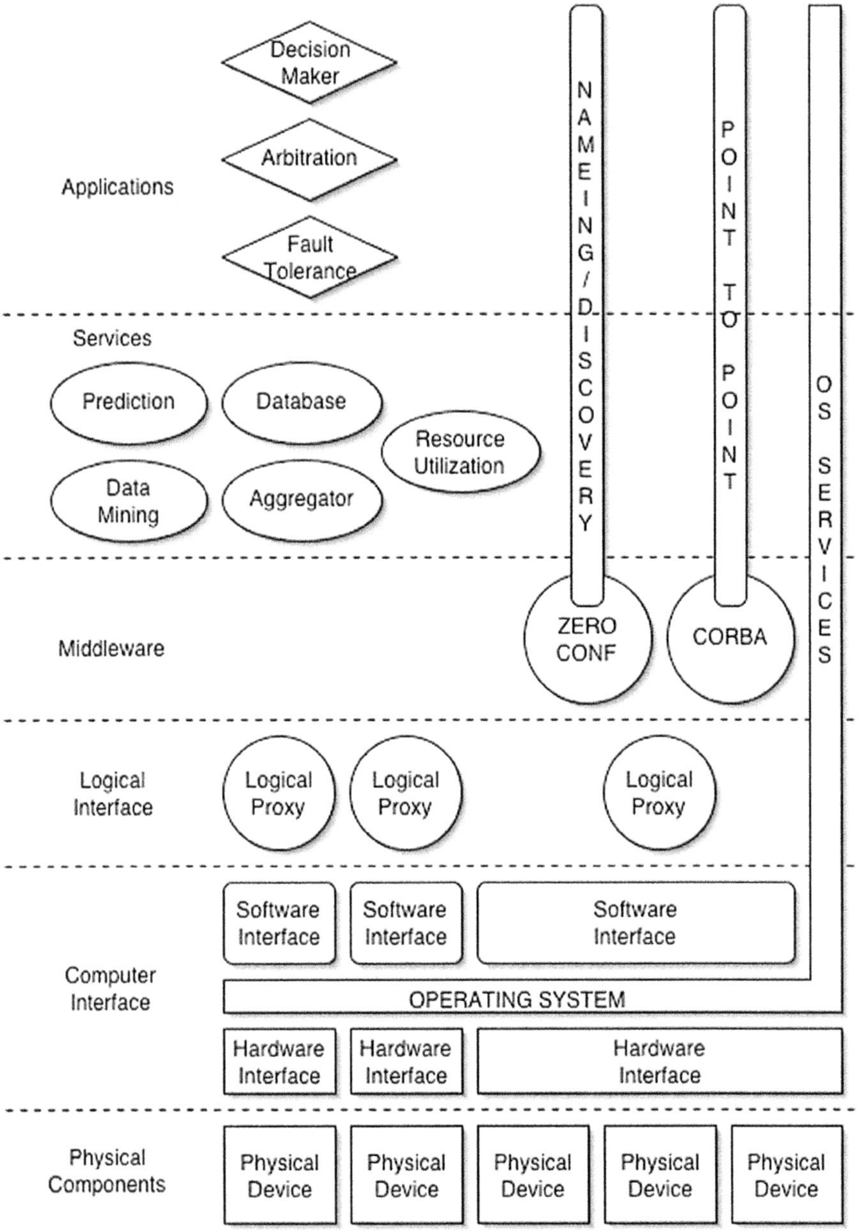

Figure 2. MavHome architecture.

(e.g., patterns, predictions). New information is presented to the decision-making applications (top layer) upon request or by prior arrangement. Action execution flows top-down. The decision action is communicated to the services layer which records the action and communicates it to the physical components. The physical layer performs the action using powerline control and other automated hardware, thus changing the state of the world and triggering a new perception.

Communication between high-level components is performed using CORBA, and each component registers its presence using zero configuration (ZeroConf) technologies. Implemented services include a PostgreSQL database that stores sensor readings, prediction components, data mining components, and logical proxy aggregators. Resource utilization services monitor current utility consumption rates and provide usage estimates and consumption queries.

MavHome is designed to optimize a number of alternative functions, but we initially focused on minimization of manual interactions with devices. The MavHome components are fully implemented and were used to automate the environments shown in Figures 3 through 5. The MavLab environment contains work areas, cubicles, a break area, a lounge, and a conference room. MavLab is automated using 54 X-10 controllers and the current state is determined using light, temperature, humidity, motion, and door/seat status sensors. The MavPad is an on-campus apartment hosting a full-time student occupant. MavPad is automated using 25 controllers and provides sensing for light, temperature, humidity, leak detection, vent position, smoke detection, CO detection, motion, and door/window/seat status sensors.

To automate our smart environment, we collect observations of manual inhabitant activities and interactions with the environment. We then mine sequential patterns from this data using a sequence mining algorithm, ED. Next, our ALZ algorithm predicts the inhabitant's upcoming actions using observed historical data. Finally, a hierarchical Markov model is created by our ProPHeT algorithm using low-level state information and high-level sequential patterns and is used to learn an action policy for the environment. Figure 6 shows how these components work together to improve the overall performance of the smart environment. In our initial study, we were able to use these software components and data collected in the smart environments to identify frequent resident behavior patters, predict sensor events, and ultimately automate 76% of the resident's interactions with the environments (Youngblood and Cook, 2007).

Our initial MavHome implementation and experiments indicated that it is possible to analyze and predict resident activities and to use this information for environment automation. This technology finds application in resident health monitoring as well. For this application, however, we see that the software algorithms could be improved by making use of timing information and temporal relationships to improve event prediction and to perform

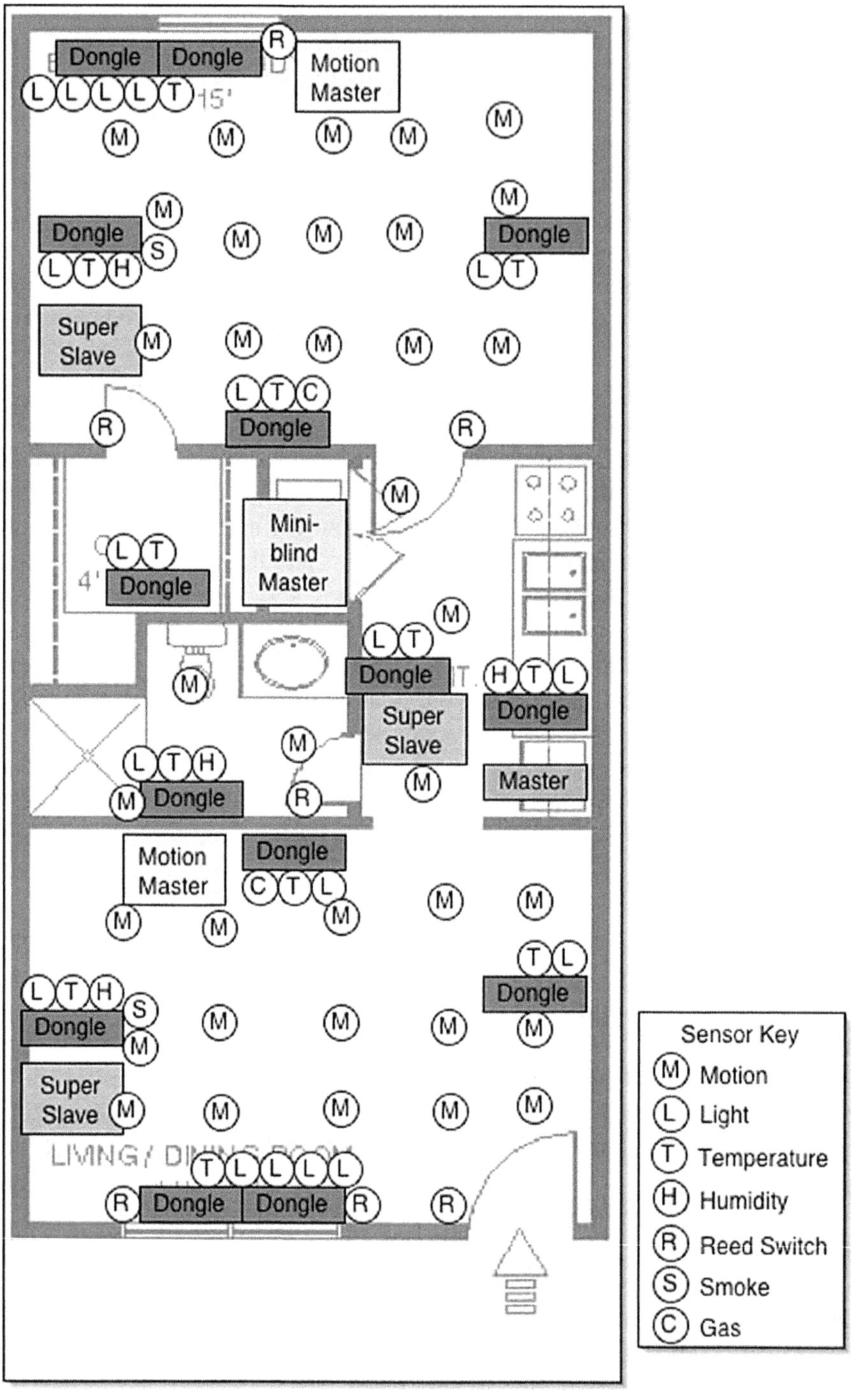

Figure 3. MavPad sensor layout.

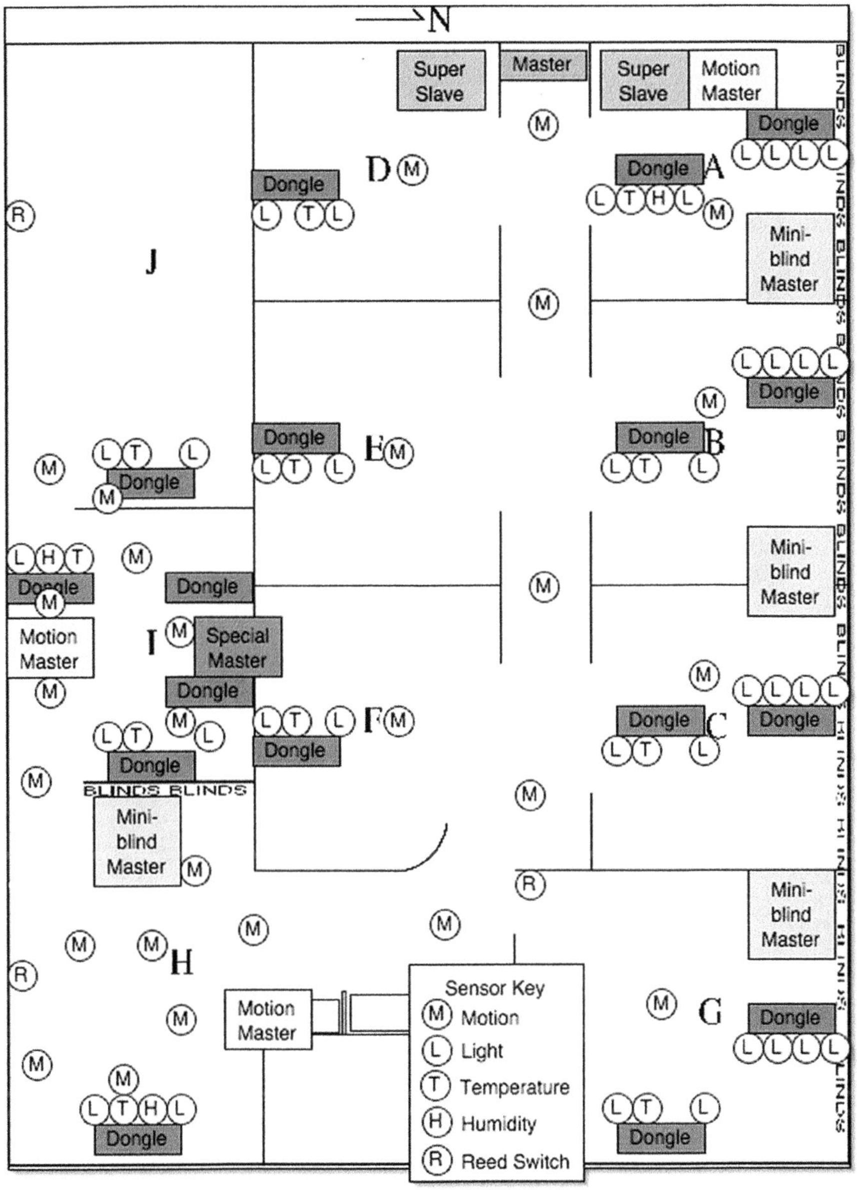

Figure 4. MavLab sensor layout.

Figure 5. MavLab rooms.

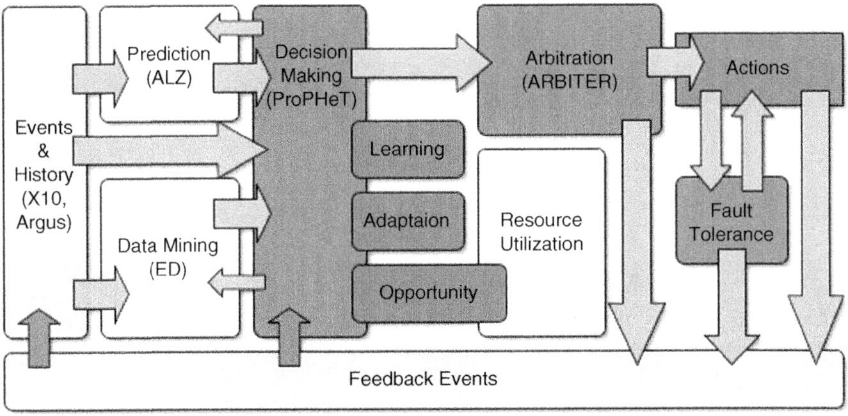

Figure 6. Integration of AI techniques into MavHome architecture.

anomaly detection. Both of these features will allow MavHome to do a more effective job of monitoring the safety of the environment and its residents. In the next section we introduce a suite of software tools designed to provide these needed features.

4. TempAl

TempAl (pronounced as "temple") is a suite of software tools which enrich smart environment applications by incorporating temporal relationship information for various applications including anomaly detection. In smart homes, the time when an event takes place is known and recorded. Our earlier MavHome algorithms did not incorporate time into its data analysis. We hypothesize that including this information would improve the performance of the smart home algorithms, which motivates our contributions of storing, representing, and analyzing timing information. The temporal nature of the data provides us with a better understanding of the nature of the data. We see that using a time series model is a common approach to reasoning about individual time-based events. However, we consider events and activities using time intervals rather than time points, which is appropriate for smart environment scenarios. As a result, we have developed methods for finding interesting temporal patterns as well as for performing anomaly detection based on these patterns.

The architecture of TempAl and its integration with the MavCore is shown in Figure 7. Raw data is read and processed by a parser to identify interval data, which is later read by a temporal relations formulation tool to identify the temporal relations. The temporal relations data is later used by the anomaly detection and event prediction components, to enhance the performance of these individual algorithms.

The objective of this study is to determine if anomalies can be effectively detected in smart home data using temporal data mining. Specifically, we introduce a temporal representation that can express frequently occurring relationships between smart environment events. We then use the observed history of events to determine the probability that a particular event should or should not occur on a given day, and report as an anomaly the presence (or absence) of highly unlikely (highly likely) events.

The need for a robust anomaly detection model is as essential as a prediction model for any intelligent smart home to function in a dynamic world. For a smart environment to perform anomaly detection, it should be capable of applying the limited experience of environmental event history to a rapidly changing environment, where event occurrences are related by temporal relations. For example, if we are monitoring the well-being of an individual in a smart home and the individual has not opened the refrigerator after he/she got out of bed as he/she normally does, this should be reported to the individual and the caregiver. Similarly, if the resident turned on the bathwater, but has not turned it off before going to bed, the resident or the caregiver should be notified, and the smart home could possibly intervene by turning off the water.

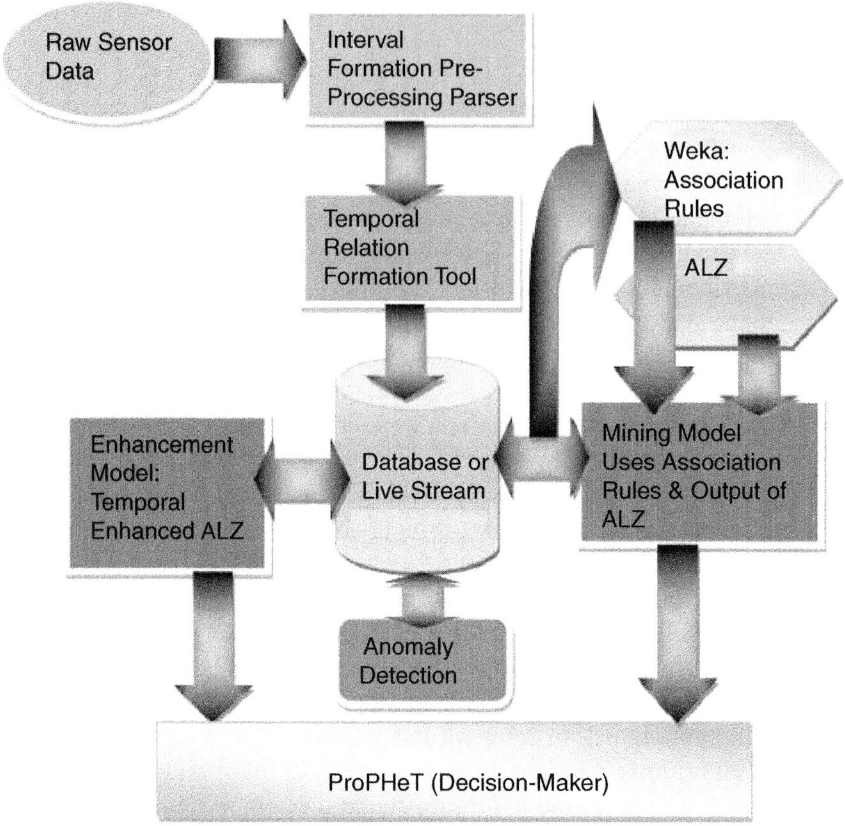

Figure 7. Architecture overview of TempAl.

4.1 Identification of Frequent Temporal Relationships

Anomaly detection is most accurate when it is based on behaviors that are frequent and predictable. As a result, we look for temporal interactions only among the most frequent activities that are observed in resident behavior. This filtering step also greatly reduces the computational cost of the algorithm. To accomplish this task, we mine the data for frequent sequential patterns using a sequence mining version of the Apriori algorithm (Agrawal and Srikant, 1995). The input to the algorithm is a file of sensor events, each tagged with a date and time, and the result is a list of frequently occurring events, which occur most frequently among the inputted file of sensor events. The pseudocode for the algorithm is given below:

```
C_k: Candidate itemset of size k
L_k: Frequent itemset of size k
L_1: {frequent items};
For (k=1; L_k ≠ ∅; k++)
do
    C_{k+1} = candidates generated from L_k;
    For each day t in dataset
    do
        Increment the count of all candidates in C_{k+1}
            that are contained in t
    end
    L_{k+1} = candidates in C_{k+1} with min_support
end
Return ∪_k L_k;
```

Next, we identify temporal relations that occur between events in these frequent sequences. The final step involves calculating the probability of a given event occurring (or not occurring), which forms the basis for anomaly detection.

4.2 Detecting Anomalies

The temporal relations that are useful for anomaly detection are the before, contains, overlaps, meets, starts, started-by, finishes, finished-by, and equals relations shown in Figure 1. Because we want to detect an anomaly as it occurs (and not after the fact), the remaining temporal relations – after, during, overlapped-by, and met-by – are not included in our anomaly detection process.

Let us focus now on how to calculate the probability that event C will occur (in this case, the start of the event interval). Evidence for this probability is based on the occurrence of other events that have a temporal relationship with C and is accumulated over all such related events. First consider the probability of C occurring given that the start of the temporal interval for event B has been detected. The formula to calculate the probability of event C based on the occurrence of event B and its temporal relationship with C is given by the equation

$$P(C|B) = (|Before(B,C)| + |Contains(B,C)| + |Overlaps(B,C)| + \\ |Meets(B,C)| + |Starts(B,C)| + |StartedBy(B,C)| + \\ |Finishes(B,C)| + |FinishedBy(B,C)| + |Equals(B,C)|)/|B|. \quad (8.1)$$

Note that the probability of B is based on the observed frequency of the observed temporal relationships between B and C as well as the number of occurrences of B in the collected event history. In this equation, we compute the probability of the occurrence of C given that B occurred using the temporal relations frequency shared between the two events. This probability count includes only those relations which aid anomaly detection. These values are added as they do not overlap and the constraints strictly enforce bounds to check that the relations are unique and thus the probability count includes the sum of their occurrences.

The previous discussion showed how to calculate the likelihood of event C given the occurrence of one other event B. Now consider the case where we want to combine evidence from multiple events that have a temporal relationship with C. In our example we have observed the start of event A and the start of event B and want to establish the likelihood of event C occurring. The combined probability is computed as

$$P(C|A \cup B) = P(C \cap (A \cup B))/P(A \cup B). \tag{8.2}$$

In this equation we calculate the probability of event C occurring (here C is the most recent event) when A and B are both frequent events and both have occurred. When both A and B occur they may also have some temporal relationships in common (i.e., a relationship that A has with B and an inverse relationship that B has with A). In these cases, one of the relationships is removed to avoid repetitive counts. An alternative computation would be $P(AB|C)$, which would be interpreted as given that C occurred, determine whether A and B are anomalies. Our approach looks at the most current observed event C and calculates evidence supporting the claim that C is an anomaly.

Such anomaly detection is useful in health monitoring. When combined with a decision maker, a smart environment could respond to a detected anomaly by reminding the resident of the needed event or automating the event. Using this second equation we can calculate the likelihood of event C occurring based on every event we have observed on a given day to that point in time. We also need to note that for the anomaly detection process we consider that each day starts with a blank slate and as the events occur new anomaly values are computed. We can similarly calculate the likelihood that an event C would not occur as $P(\neg C) = 1 - P(C)$. Finally, we calculate the anomaly value of event C using the equation $Anomaly_C = 1 - P(C)$.

Note that if the event has an anomaly probability approaching 1 and the event occurred, this is considered an anomaly. Similarly, if the probability is close to 0 and the event does not occur then it should also be considered an anomaly. The point at which these anomalies are considered surprising enough to be reported is based somewhat on the data itself (Noble and Cook, 2003). If the probability of an event is based on the occurrence of other events

Timestamp	Sensor State	Sensor ID
...		
3/3/2003 11:18:00 AM	OFF	E16
3/3/2003 11:23:00 AM	ON	G12
3/3/2003 11:23:00 AM	ON	G11
3/3/2003 11:24:00 AM	OFF	G12
3/3/2003 11:24:00 AM	OFF	G11
3/3/2003 11:24:00 AM	ON	G13
3/3/2003 11:33:00 AM	ON	E16
3/3/2003 11:34:00 AM	ON	D16
3/3/2003 11:34:00 AM	OFF	E16
...		

which themselves rarely occur, then the evidence supporting the occurrence of the event is not as strong. In this case, if the event has a low probability yet does occur, it should be considered less anomalous than if the supporting evidence itself appears with great frequency. Consistent with this theory, we calculate the mean and standard deviation of event frequencies over the set of frequent events in the resident's action history. An event (or, conversely, the absence of an event) is reported as an anomaly if it does (does not) occur and its anomaly value is greater (lesser) than the mean probability + 2 standard deviations (or mean -2 standard deviations). Two standard deviations away from the mean accounts for roughly 95%, so any value which falls out of this population would be reported as an anomaly.

To illustrate the process, we start with a sample of raw data collected in the MavLab environment:

Next, we identify (start, end) time intervals that are associated with the events. Here is a sample of the time intervals that are associated with the raw data:

Date	Sensor ID	Start Time	End Time
...			
03/02/2003	G11	01:44:00	01:48:00
03/02/2003	G13	04:06:00	01:48:00
03/03/2003	E16	11:18:00	11:34:00
03/03/2003	G12	11:23:00	11:24:00
...			

Once the time intervals are established we discover temporal relations that frequently exist among these events, such as the ones shown here:

Time	Sensor ID	Temporal Relation	Sensor ID
...			
3/3/2003 12:00:00 AM	G12	DURING	E16
3/3/2003 12:00:00 AM	E16	BEFORE	I14
3/2/2003 12:00:00 AM	G11	FINISHESBY	G11
4/2/2003 12:00:00 AM	J10	STARTSBY	J12
...			

The frequencies of these relationships are tabulated and used as the basis for calculating anomaly values each time a new event is observed. When an event occurs which has a sufficiently high anomaly value, the event is reported as an anomaly.

5. Experimental Findings

To validate our TempAl anomaly detection algorithm, we apply it to real and synthetic smart home data. Table 1 summarizes features of the data sets used for these experiments. The real data represents raw sensor data collected for 60 days in the MavLab environment with a volunteer resident. The synthetic data represents instances of a predefined set of activities. In the synthetic data we have injected anomalous events to see if TempAl will catch these events and label them as anomalies.

To test our algorithms we train the models using 59 days of sensor data, then test the model on a single day of events. Table 2 shows the anomaly values that are calculated for the 8 observed events in the real data, and Table 3 shows the anomaly values for the 17 observed events in the synthetic data. The values are visualized in Figure 8.

Based upon a visual inspection of the data we see that the anomaly detection algorithm performed well – all of the expected anomalies were detected and no false positives were reported. In the real data no anomalies are reported, which is consistent with the nature of the data. This result reflects the fact that anomalies should be, and are in fact, rare. We see that the TempAl algorithms are robust in this case and do not report false positives.

Table 1. Parameter settings for experiments.

Data sets	Number of Days	Number of Events	Number of Identified intervals	Data set size (KB)
Real	60	17	1623	104
Synthetic	60	8	1729	106

Table 2. Anomaly detection in real data test set.

Frequent Event ID (in chronological order)	Frequent event	Event probability	Anomaly value	Anomaly detected
1	J10	0.45	0.55	No
2	J11	0.32	0.68	No
3	A11	0.33	0.67	No
4	A15	0.24	0.76	No
5	A11	0.23	0.77	No
6	A15	0.22	0.78	No
7	I11	0.27	0.73	No
8	I14	0.34	0.66	No
Anomaly mean			0.7	
Anomaly standard deviation			0.07	
Anomaly threshold			0.84	

Table 3. Anomaly detection in synthetic data test set.

Frequent Event ID (in chronological order)	Frequent event	Event probability	Anomaly value	Anomaly detected
1	Lamp	0.30	0.70	No
2	Lamp	0.23	0.77	No
3	Lamp	0.01	0.99	Yes
4	Fan	0.32	0.68	No
5	Cooker	0.29	0.71	No
6	Lamp	0.45	0.55	No
7	Lamp	0.23	0.77	No
8	Lamp	0.01	0.99	Yes
9	Lamp	0.23	0.77	No
10	Fan	0.30	0.70	No
11	Cooker	0.34	0.66	No
12	Lamp	0.33	0.67	No
13	Lamp	0.20	0.80	No
14	Lamp	0.02	0.98	No
15	Lamp	0.00	1.00	Yes
16	Fan	0.34	0.66	No
17	Cooker	0.42	0.58	No
Anomaly mean			0.76	
Anomaly standard deviation			0.14	
Anomaly threshold			0.99	

The experimental results summarized here provide evidence that our algorithm is capable of identifying anomalous events based on temporal relationship information. The results applied to real data bring insights into the

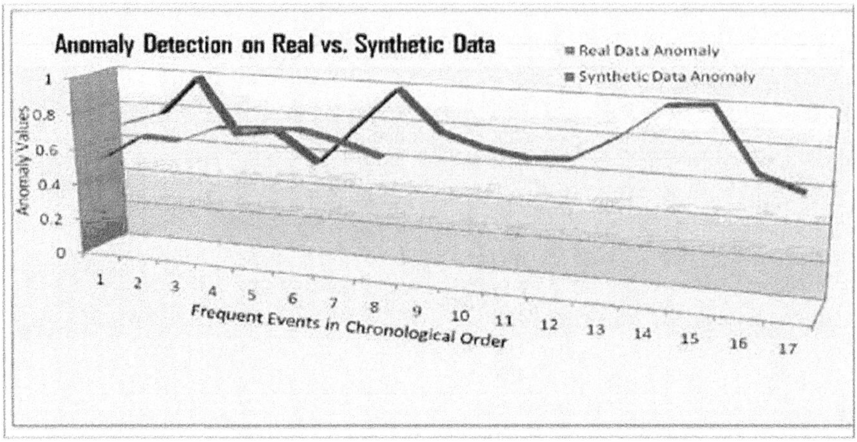

Figure 8. Visualization of calculated anomaly values for real and synthetic data. The spikes in the graph are events which are flagged as anomalies.

activities that were being performed in the MavLab setting. In both cases the anomalies would be reported to the resident and possibly the caregiver. The caregiver could respond according to the health-critical nature of the anomaly and any additional information he/she possesses.

An extended application of anomaly detection is its use for reminder assistance. If the resident queries the algorithm for the next routine activity, the expected activity or activities with the greatest probability can be provided. Similarly, if an anomaly is detected, the smart environment can first initiate contact with the resident and provide a reminder of the activity that is usually performed at that time. Autominder (Pollack et al., 2003) is an example of a reminder system that has already been developed for this purpose using techniques such as dynamic programming and Bayesian learning. Unlike our approach, Autominder does not base its generated reminders on a model of behavior that is learned from actual observed events.

6. Conclusion and Future Work

Temporal reasoning enhances smart environments algorithms by incorporating learned information about temporal relationships between events in the environment. Based on our study, we conclude that the use of temporal relations provides us with an effective new approach for anomaly detection. We tested TempAl on relatively small data sets, but will next target larger data sets with real data collected over a 6-month time span.

Another major enhancement to this work would be to consider an interval analysis of intermediate device states. Intermediate states are those which exist between an ON and OFF state. For example, a lamp controlled by a dimmer switch will have a number of light levels supported by the dimmer, which form the intermediate states. Identifying time intervals for events changing intermediate states would be a challenge for TempAl but would provide more refined information to the algorithms.

In addition, we would like to investigate other techniques for identifying the anomaly threshold function. Other future works can focus on finding better fusion techniques to enhance existing anomaly detection algorithms using temporal relationship information.

While making sense of sensor data can be challenging for smart environment algorithms, the problem is made even more complex when the environment houses more than one resident (Jakkula et al., 2007). To aid the capabilities of our temporal data mining algorithms and to reveal the complexities of multi-resident spaces, an entity discovery tool is needed. Enriching the raw data set provided by the smart environment gives the knowledge discovery tools more information to use during the mining process. This comes in the form of an entity (in this case, resident) identification number that is attached to each event, matching events to entities. Thus, using temporal activity models to identify patterns and associate these patterns to behavior models for entity identification and resident profiling is a direction we are currently pursuing.

Agents in dynamic environments have to deal with changes over time. Enhancing TempAl to detect changes in temporal relationship frequencies and to use this information would be a good future direction of this work.

Acknowledgments This work was partially supported by National Science Foundation grant IIS-0121297.

References

Agrawal, R. and Srikant, R. (1995). Mining Sequential Patterns. In *Proceedings of the International Conference on Data Engineering*, pages 3–14.

Allen, J. F. and Ferguson, G. (1994). Actions and Events in Interval Temporal Logic. *Journal of Logic and Computation*, 4(5):531–579.

Dekhtyar, A., Ross, R., and Subrahmanian, V. S. (2001). Probabilistic Temporal Databases, I: Algebra. *ACM Transactions on Database Systems*, 26(1):41–95.

Gottfried, B., Guesgen, H. W., and Hubner, S. (2006). Spatiotemporal Reasoning for Smart Homes. In *Designing Smart Homes*, pages 16–34. Springer, Berlin / Heidelberg.

Gross, J. (August 14, 2007). A Grass-Roots Effort to Grow Old at Home. In *The New York Times*.

Jakkula, V., Crandall, A., and Cook, D. J. (2007). Knowledge Discovery in Entity Based Smart Environment Resident Data Using Temporal Relations

Based Data Mining. In *Proceedings of the ICDM Workshop on Spatial and Spatio-Temporal Data Mining.*

Lanspery, S., Callahan, Jr., J. J., Miller, J. R., and Hyde, J. (1997). Introduction: Staying Put. In *Staying Put: Adapting the Places Instead of the People*, pages 1–22. Baywood Publishing Company, Amityville, NY.

Morchen, F. (2006). Algorithms for Time Series Mining. In *Proceedings of the 12th ACM SIGKDD International Conference on Knowledge Discovery and Data Mining*, pages 668–673.

Noble, C. and Cook, D. J. (2003). Graph-Based Anomaly Detection. In *Proceedings of the 9th ACM SIGKDD International Conference on Knowledge Discovery and Data Mining.*

Pollack, M., Brown, L., Colbry, D., McCarthy, C., Orosz, C., Peintner, B., Ramakrishnan, S., and Tsamardinos, I. (2003). Autominder: An Intelligent Cognitive Orthotic System for People with Memory Impairment. *Robotics and Autonomous Systems*, 44(3-4):273–282.

Ryabov, V. and Puuronen, S. (2001). Probabilistic Reasoning about Uncertain Relations Between Temporal Points. In *Proceedings of the 8th International Symposium on Temporal Representation and Reasoning*, pages 35–40.

Wang, H. (2006). Digital Home Health – A Primer. In *Parks Associates.*

Worboys, M. F. and Duckham, M. (2002). Integrating Spatio-Thematic Information. In *Proceedings of the International Conference on Geographic Information Science*, pages 346–361.

Youngblood, G. M. and Cook, D. J. (2007). Data Mining for Hierarchical Model Creation. *IEEE Transactions on Systems, Man, and Cybernetics, Part C*, 37(4):1–12.

Youngblood, G. M., Holder, L. B., and Cook, D. J. (2005). Managing Adaptive Versatile Environments. *Journal of Pervasive and Mobile Computing*, 1(4):373–403.

Chapter 9

FAULT-RESILIENT PERVASIVE SERVICE COMPOSITION

Hen-I Yang*, Raja Bose*, Abdelsalam (Sumi) Helal

Department of Computer and Information Science and Engineering, University of Florida, Gainesville, Florida USA

{hyang, rbose, helal}@cise.ufl.edu

**Joint First Authors*

Jinchun Xia, Carl K. Chang

Department of Computer Science, Iowa State University, Ames, Iowa, USA

{jxia, chang} @iastate.edu

Abstract Service-oriented architecture (SOA) promises an elegant model that can easily handle dynamic and heterogeneous environments such as pervasive computing systems. However, in reality, frequent failures of resource-poor and low-cost sensors greatly diminish the guarantees on reliability and availability expected from SOA. To provide a framework for building fault-resilient, service-oriented pervasive computing systems, we present a solution that combines a virtual sensor framework with WS-Pro/ASCT, a service composition mechanism. The use of virtual sensors enhances the availability of services, while the service composition solution ensures that the system can efficiently adapt to changes and failures in the environment. This approach also includes a novel probe-based monitoring technique for proactive collection of performance data and a Finite Population Queuing System Petri Net (FPQSPN) for modeling the performance of composed services.

Keywords: Fault-resilient pervasive services; Pervasive service composition; Service composition optimization; Pervasive computing; Virtual sensors.

1. Introduction

Service-oriented architecture (SOA) has established itself as a "prevailing software engineering practice" in recent years (McCoy and Natis, 2003). This popularity extends to the domain of pervasive computing as its characteristics

W. Minker et al. (eds.), *Advanced Intelligent Environments*,
DOI 10.1007/978-0-387-76485-6_9,

of loose-coupling, statelessness, and platform independence make it an ideal candidate for integrating pervasive devices. While the semantics of SOA are being standardized, its use in pervasive computing is the subject of extensive research and experimentation. In fact, SOA-based pervasive computing systems are fast becoming a reality with the introduction of technology such as the Atlas Platform (King et al., 2006). Introduced in 2006, it is the world's first commercially available service-oriented sensor and actuator platform and provides basic building blocks for creating pervasive computing systems using SOA (Bose et al., 2006).

However, in spite of all the promises offered by favorable characteristics of SOA for coping with dynamic and heterogeneous environments, one should not forget that underneath all the nice wrappings of highly reliable and self-integrating services, the actual data sources are mass-deployed, low-end sensors that are poor in terms of resources available and inherently unreliable, both because of the large number of entities deployed and because of the common choice of employing low-cost components with little guarantee of their quality.

Unlike web services which are mostly hosted by high-end servers and data centers, where service failures are rare, pervasive services need to embrace service failures as the norm. For these services to work properly and reliably, mechanisms need to be in place to improve their availability and assess the quality of their data so that necessary adjustments can be made.

Our proposed solution for building fault-resilient pervasive computing systems consists of two parts. The first part is the virtual sensor framework (Bose et al., 2007) which improves the availability of basic component services. The second part consists of an architecture for performing service composition that can efficiently model, monitor, and re-plan this process. In this architecture, WS-Pro (Xia, 2006; Xia and Chang, 2006), the probe-based web service composition mechanism is adjusted to support the Abstract Service Composition Template (ASCT), a template-based service composition scheme for providing generic solutions for high-performance pervasive service composition.

To create a comprehensive solution, these two parts have to work hand in hand during the entire life cycle of pervasive services. During the design stage, programmers examine the functional requirements and the type of physical sensors available to create virtual sensor services, which can then be used to match against the specifications in ASCT. During the execution stage, the compensation provided by virtual sensors provides the first line of defense against sensor failures. However, if the failures are widespread or occur within service components of higher abstraction, the WS-Pro/ASCT mechanism kicks in to identify replacement services for the failed ones.

The rest of this chapter is organized as follows. In Section 2, we provide a classification of basic pervasive service and examine the main requirements for building fault-resilient pervasive services. In Section 3, we present the first part of the solution, which is the concept of virtual sensors. In Section 4 we discuss the performance management of pervasive service composition, and then present the second part of the solution, which is the WS-Pro/ASCT approach for efficient service re-planning and composition. In Section 5, we evaluate the performance of both mechanisms using realistic examples and demonstrate how they enhance the reliability, availability, and adaptability of pervasive services. Section 6 describes how both parts come together as a comprehensive solution, focusing on the software engineering aspect of integrating virtual sensors with ASCT, as well as the compensation and adjustment mechanisms at runtime. We present the related work in Section 7, followed by future work and conclusions in Section 8.

2. A Brief Primer on Pervasive Services

We begin this section by providing a classification of basic pervasive services and describe their respective roles. Then, we discuss the major requirements for building fault-resilient pervasive services.

2.1 Classification of Basic Pervasive Services

Pervasive services can be categorized into three types of abstract service elements based on their functionalities and the roles they play. As depicted in Figure 1, a typical end-to-end service consists of a context provider, context processor, and information deliverer.

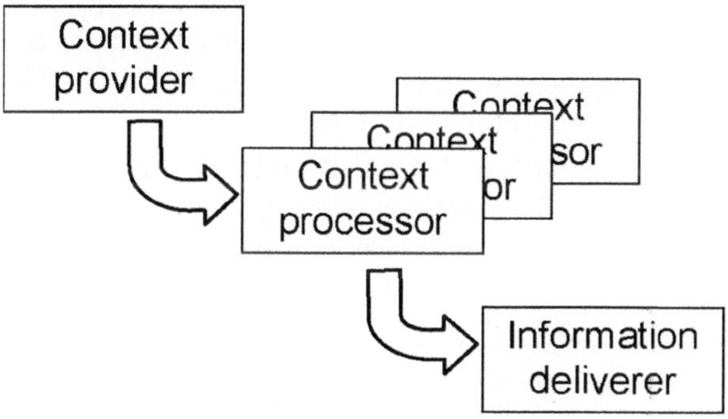

Figure 1. Composition of a typical end-to-end pervasive service.

A context provider is an abstract service element that retrieves context-specific data from sensors. In other words, a context provider is a wrapper for input sensing components for detecting a specific context. In addition, each context provider has internal test functions for monitoring the health and data quality of its member components. For example, a weather monitor as a wrapper of real or virtual sensors can obtain the current temperature, humidity, and wind velocity outside a house.

The context processor is the major functional component that deals with data from context providers and produces meaningful context-based information for information deliverers.

The information deliverer is also a wrapper of a specific hardware component, such as a monitor, a printer, or an emergency alarm, which is used to present the information generated by the context processor. An information deliverer includes a transcoding feature that transforms the information created by the context processor into a more appropriate format. For example, a printing deliverer creates a PDF file based on the information fed from context processors and sends it to a printer.

Context providers and information deliverers are services bound to physical devices and they encounter more performance problems than context processors. The problem may exist either in the hardware layer (such as device connectivity failure) or in the service layer (protocols, service failures, etc.). We refer to these two types of errors as data errors and configuration errors, respectively.

2.2 Requirements for Building Fault-Resilient Pervasive Services

To build a fault-resilient pervasive computing system, we look for measures that can drastically improve the availability and adaptability of services. Here availability is tightly related to data errors discussed above and adaptability issue addresses configuration errors at the application level. To enhance the availability, we deploy multiple or even redundant sensors in the same vicinity so that a certain number of failed sensors can be algorithmically compensated for by utilizing readings from their neighbors. This is enabled by the virtual sensor mechanism described in Section 3. To ensure that services can adapt to a changing environment, we designed a framework that allows efficient dynamic re-planning so that a failed service can be quickly replaced with a substitute service before it becomes responsible for bringing down the entire service chain. This framework is called WS-Pro and is described in Section 4.

The most intuitive way to represent sensors in SOA is to simply wrap each physical sensor as an individual service implemented in software. However, by bundling sensors with similar functionalities in close proximity, and representing them collectively as a single service (virtual sensor), we not only improve its resiliency against failure but also provide the means to gauge the quality of data collected.

The WS-Pro framework is responsible for service composition at the application level. It aims at fixing configuration errors and improving the overall service computing performance. SOA allows the reuse and integration of existing software components via service composition – the most critical process of service-oriented computing. The nature of SOA allows this service composition to be either static or dynamic. Compared to standard web services featuring business logic, service-oriented pervasive computing is much more domain specific and poses additional challenges such as lack of flexibility in re-planning due to context and hardware constraints. Since failure is the norm in pervasive computing, dynamic service composition and re-planning is required to ensure uninterrupted services.

Not only do services need to be composed dynamically amid failures and changes, but the composition and planning process has to be performed efficiently. Performance has been identified as the most critical attribute of quality by many researchers. There is an overwhelming demand from industry for robust and verifiable web services with high performance (Ludwig, 2003; Srivastava and Koehler, 2003). Our experience in the Gator Tech Smart House (Helal et al., 2005) further confirms that poor performance in service adaptation can seriously hinder efforts to adopt SOA as the solution for pervasive computing.

3. Virtual Sensors

A virtual sensor is a software entity, representing a group of sensors annotated with associated knowledge, which enables it to provide services beyond the capabilities of any of its individual components. Virtual sensors may be composed of a group of physical sensors or other virtual sensors.

3.1 Classes of Virtual Sensors

Virtual sensors can be classified into one of the following three categories: *singleton, basic,* and *derived virtual sensor.* A singleton virtual sensor represents a single physical sensor, whereas a basic virtual sensor is composed of a group of singleton virtual sensors of the same type, and a derived virtual sensor is composed of a group of basic and/or other derived virtual sensors of heterogeneous types.

Virtual sensors enhance the availability of the sensor services and the reliability of the overall sensor network. They provide certain basic guarantees of functionality and are capable of estimating the reliability of data originating from the source sensors. They also have mechanisms for recovering from failures of multiple physical sensors and detecting degradation in their performance.

3.2 The Virtual Sensor Framework

The virtual sensor framework is responsible for managing the life cycle of all the virtual sensors running inside the service environment. The architecture of the framework is shown in Figure 2.

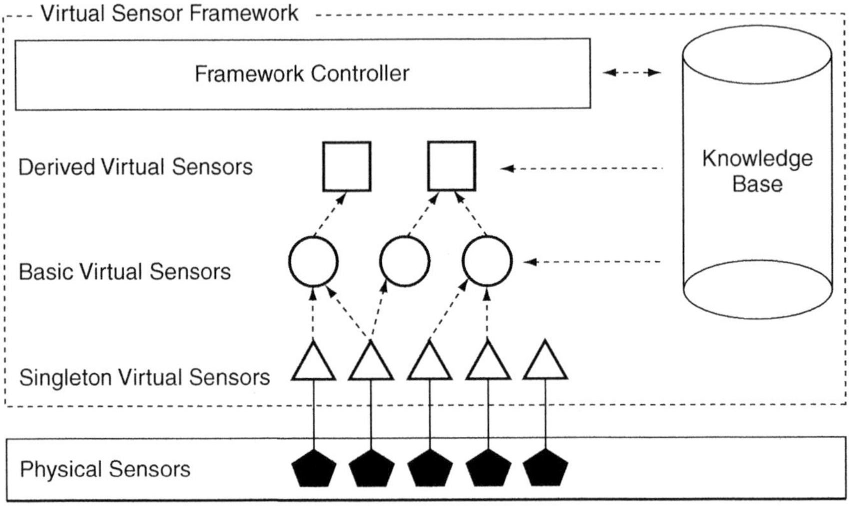

Figure 2. Architecture of the virtual sensor framework.

The knowledge base manages information related to virtual sensors, such as reliability and availability parameters, sensor model definition for each type of virtual sensor, and phenomena definitions that describe the types of virtual sensor capable of detecting a certain phenomenon.

The framework controller is responsible for receiving queries from applications and then determining which virtual sensors need to be created with the help of the knowledge base. Dynamic creation and organization of virtual sensors in the framework allow the sensor network to fulfill queries on phenomenon detection despite failures and changes. When a query is issued to detect a certain phenomenon, the framework controller identifies the type of virtual sensor required from the knowledge base. It then instantiates the target virtual sensor based on the sensor model definition. If the target virtual sensor relies on other virtual sensors as data sources, it subscribes to their data if the source virtual sensors already exist, otherwise it instantiates the necessary data sources.

3.3 Monitoring Quality of Basic Virtual Sensors

Within each basic virtual sensor, we constantly monitor the correlated behavior of component singleton virtual sensors and record whether the differences among their readings are within an acceptable margin ϵ. This record is used to calculate the weight factor W $\in$ [0, 1] if particular singleton virtual sensors fail and the readings from other singletons are used to approximate their readings. The similarity factor between two sensors is the probability that the differences between their data readings are within ϵ. The more similar the readings are between a pair of singleton virtual sensors, the larger the similarity factor, and therefore larger the value of weight W assigned to one when the other dies. This compensation mechanism mitigates the effect of failed sensors and enhances the overall availability of sensor data.

A Virtual Sensor Quality Indicator (VSQI) is associated with each basic virtual sensor, which measures the reliability and confidence level of its data. The formula for computing VSQI is given as

$$VSQI_{BVS} = \frac{(num\_of\_sensor\_alive + \sum_{s \in all\_failed\_sensors}(1 - e^{\frac{-t}{a}})W_s)}{total\_num\_of\_sensors} \tag{9.1}$$

where W_s is the weight factor W associated with sensor s, t is the time elapsed since the sensor network started, and a is a constant greater than 0 whose value depends on the sensor and the environment in which it is deployed.

VSQI is measured on a scale of 0–1. A basic virtual sensor will have a VSQI value of 1 if all its member sensors are functioning properly. However, this VSQI formula assumes that all physical sensors either function properly or die. The formula adjusts the penalty caused by the discrepancies between the failed sensor and the sensor used to compensate for the failure. The term $1 - e^{\frac{-t}{a}}$ is used to factor in the confidence level associated with a particular weight factor W_s based on the length of observation. If a W_s value is calculated using data collected over a longer duration, $1 - e^{\frac{-t}{a}}$ is larger, which means it carries more weight than when it is calculated over shorter lengths of observation. To ensure data quality and service reliability, we define $VSQI_T$ as the threshold of reliable sensor quality. Any sensor readings with $VSQI < VSQI_T$ are deemed unreliable, and the source virtual sensor is taken off-line to preserve the integrity of the system. Readers interested in learning more about the virtual sensor framework and the VSQI formula are encouraged to refer to Bose et al. (2007).

The results from a simplified simulation, as shown in Figure 3, demonstrates how the adoption of virtual sensors significantly increases the availability and reliability of sensor data in the face of sensor failure. The degree of similarity in behavior of readings among the sensors, which is heavily influenced by

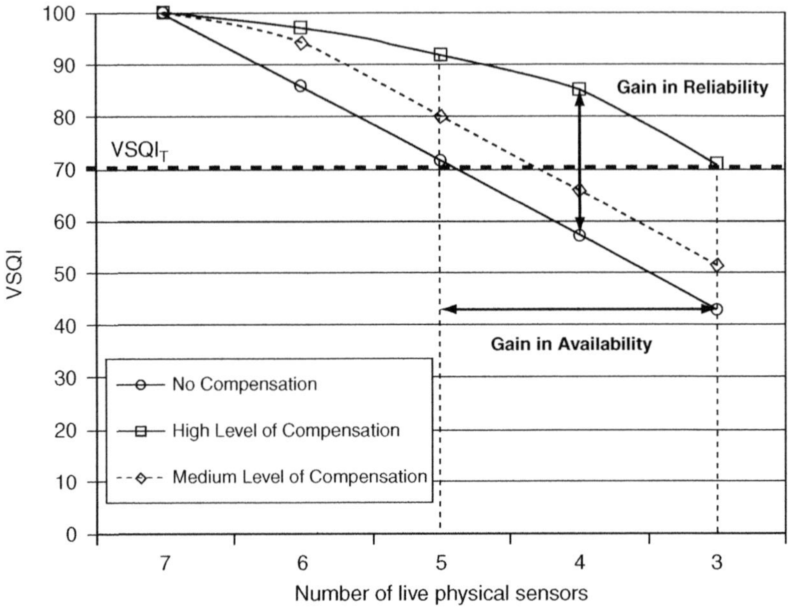

Figure 3. Comparison between different levels of compensation.

factors such as pattern of sensor deployment and the environment where they are operating in, decides a virtual sensor's effectiveness in improving fault resiliency. When the readings are highly similar, even with small number of sensors, we can observe the availability of the sensor network is doubled and furthermore, there is an average gain of 30% in reliability, as compared to the case where there is no compensation.

3.4 Monitoring Quality of Derived Virtual Sensors

A derived virtual sensor may be composed of numerous heterogeneous basic or derived virtual sensors. Hence, monitoring the performance of a derived virtual sensor requires a slightly different approach than a basic virtual sensor.

The $VSQI_{BVS}$ formula gives us the probability that the compensated output of a basic virtual sensor matches the expected output, if it was fully functional. Following the same trend of thought, we define the VSQI associated with a derived virtual sensor as the product of the VSQIs of its member basic and derived virtual sensors:

$$VSQI_{DVS} = \prod_{i=1}^{n} VSQI_{DVS_i} \times \prod_{j=1}^{m} VSQI_{BVS_j}. \tag{9.2}$$

4. Efficient Pervasive Service Composition

Service composition has been well studied by the web service community, and the widespread adoption of SOA in pervasive computing inspires researchers to examine if the techniques designed for web services are equally applicable to pervasive services. However, there exist subtle but critical differences between the two. For instance, in web service composition it is assumed that the underlying Internet infrastructure is universal so that services can always be discovered and composed regardless of the differences in platform or communication medium used. The pervasive services, however, are tightly bound to heterogeneous hardware platforms and communication mediums, making their composition different. For example, the discovery of Bluetooth services is limited by the range of the Bluetooth device. This limitation imposes additional challenges toward composition of pervasive services.

4.1 Performance Management of Pervasive Service Composition

Hummel identifies fault tolerance as a crucial issue in pervasive services (Hummel, 2006) and points out the importance of pervasive services being able to react to dynamic changes in time. Our experience in the Gator Tech Smart House also agrees with this assessment. Different smart devices are represented as collaborating services in SOA. However, just because they work properly in collaboration does not guarantee satisfactory service to users, if they fail to deliver on time. In any typical pervasive computing environment such as a smart home, the concern for safety and security is very real and the services addressing these issues have to be delivered promptly. In addition, users usually expect such environments to respond in a reasonably short period of time. Therefore timely delivery of services is critical. By optimizing the reconfiguration and re-composition of pervasive services, we improve the fault tolerance as well as user satisfaction.

Performance management of service composition is an ongoing research area in web services and the two main issues are an active performance monitoring technique and a more efficient mechanism for service composition. Pervasive services, on the other hand, exhibit cross-organization and cross-location properties which invalidate many existing performance analytic techniques. The dynamicity of the execution environment exemplified by the frequent disappearance and re-emergence of component services delays verification and testing from design and implementation stages to the post-deployment stage. Runtime monitoring is the best solution in assessing properties of services when traditional verification techniques are insufficient, as shown in Baresi and Guinea (2005).

Unfortunately, most existing monitoring techniques are passive considering their measurements are all tightly coupled with real business process execution. When a service composer requests for performance references before committing the composition, existing local monitoring techniques cannot provide any help if historical data is lacking. Therefore, a more active technique to monitor the runtime performance status of services and provide timely support for dynamic service composition is required. Zeng et al. suggested the use of probe as one possible solution (Zeng, et al., 2004). Our solution, WS-Pro, employs a testing-based probe for active monitoring.

4.2 WS-Pro: A Service Composition Engine

WS-Pro was developed to model and optimize system performance in standard web service infrastructure. In the WS-Pro framework, the business process, represented in WS-BPEL (OASIS, 2007), is transformed into a Petri net. Using the generated Petri net model, we can verify the correctness of service composition. Meanwhile, performance metrics of available component services are derived from historical data (through log/audit system) or runtime measurements (through the probe). These performance values are used to augment the Petri net generated from the WS-BPEL model.

We designed a composition algorithm based on the stochastic Petri net that computes the best execution plan which will optimize the performance of the composite service. We are concerned about not only the performance of the composite service itself but also the performance of the composition process. Therefore, we designed the composition algorithm to be two phased. In the first phase, called off-line phase, we compress the Petri net, generate its reachability graph, and remove all the loops in the reachability graph using statistical means. Then in the second phase namely, runtime computation, we use the Djikstra's algorithm to compute the final execution plan from the unfolded reachability graph. Since generating the reachability graph and handling loops is responsible for majority of the total computation, by removing all the loops from the reachability graph, we move most of the computation to the off-line phase, thereby improving runtime performance. We also designed a re-planning algorithm to handle performance exceptions during service execution.

The architecture for applying WS-Pro in pervasive service composition is illustrated in Figure 4. In this architecture, there is a testing-based probe that actively collects runtime performance status of component services. It supports

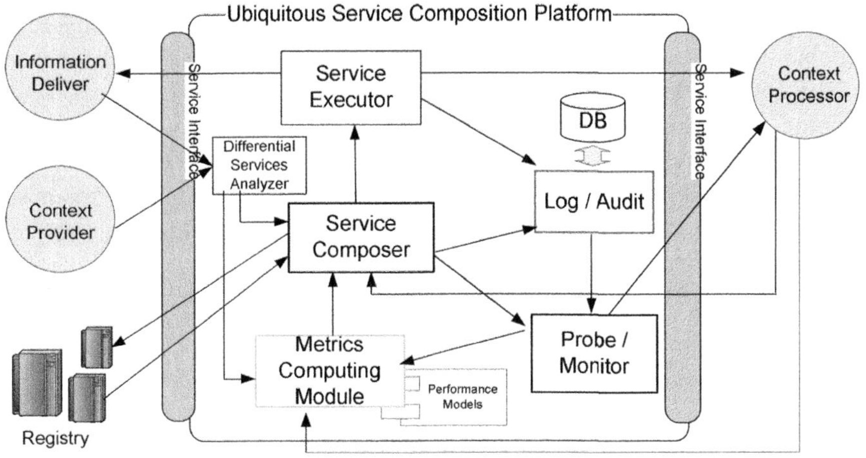

Figure 4. Overview of pervasive service composition.

runtime service composition and composition re-planning. As Figure 4 demonstrates, the service composer transforms a business process into an analytic model, which is a Petri net. The metrics computing module calculates performance values. With input from the computing module, the service composer generates an execution plan having optimal performance. This execution plan is run by the service executor. The monitor observes the service execution and whenever it captures performance exceptions, it asks the composer to re-plan the composition.

4.3 Abstract Service Composition Template (ASCT)

In web services, WS-BPEL is the language to describe a business process that represents how component services are composed together. Similarly, an Abstract Service Composition Template (ASCT) is a meta-level business process used to describe the sequential process of critical functions to deliver a composite service. The purpose of defining an ASCT is to remove the limitations of service interface description so that more services can be considered as alternatives to each other.

A similar approach using service templates is shown in Yamato et al. (2006). An ASCT consists of a critical flow of a service using abstract service elements that perform critical functions. Based on the abstract service elements in an ASCT, composite services can be materialized by searching and selecting appropriate actual service instances such as a context provider with assorted virtual sensors. The authors implemented a composition engine, evaluated its performance in terms of processing time and memory usage, and discussed the suitable size of categories and the number of service elements in each category.

The novelty of this approach is the introduction of an abstract layer for describing composite services and the support for automatic composition. As a result, scenarios of a composite service can be easily described without considering strict and detailed interface description of the intended composite services. Furthermore, this makes it possible for users to create their own services or customize existing ones.

4.4 WS-Pro with ASCT Approach

As described earlier, WS-Pro was originally designed for dynamic web service composition. However, it does not adapt well to pervasive services because of two major reasons. First, dynamic composition is important in web services because there is always a large pool of candidate component services. With pervasive computing, however, the options are often much more limited in terms of functionality, making WS-Pro unsuitable for service composition. Second, each device in pervasive computing, even low-level ones such as physical sensors, is represented as atomic services, resulting in a huge number of nodes that need to be modeled. The Generalized Stochastic Petri Net (GSPN), which is used in WS-Pro, does not scale well and hence cannot serve as an appropriate modeling tool.

To address the first problem, we introduced the notion of ASCT. By using abstract service elements instead of well-defined Web Services Definition Language (WSDL) interfaces, different devices with similar functions (for example, different presentation devices such as monitor, printer, speaker) can be discovered and composed using the same service discovery queries. This approach also allows similar services to be considered as alternative candidates. Therefore, ASCT eliminates the problem of insufficient candidate pool at the actual service instance level when applying WS-Pro in pervasive computing environments. In addition, an ASCT can further reduce the size of the Petri net that models the service composition.

The second problem can be mitigated by replacing GSPN with a Finite Population Queuing System Petri Net (FPQSPN) (Capek, 2001). FPQSPN extends the Petri net by introducing "shared places and transitions", hence greatly reducing its size. GSPN only uses exponential distribution on transitions in order to preserve time independence. However, FPQSPN allows users to use other probability distributions which extend the GSPN's stochastic modeling capability. A Finite Population Queuing System Petri Net (FPQSPN) is formally defined using a 9-tuple $(P, T, Pre, Post, M_0, s_o, t, tt, k)$ such that

- P is a finite and non-empty set of places
- T is a finite and non-empty set of transitions
- Pre is an input function, called precondition matrix of size $(|P|, |T|)$
- Post is an output function, called post-condition matrix of size $(|P|, |T|)$
- $M_0 : P(R)\{1, 2, 3, ...\}$ is an initial marking
- $t : T(R)\ t \in R+$ is the time associated to transition
- $tt : T(R)\ tt$ is the type of time associated to transition
- $s_o : T_i : T_i \in T, P_i : P_i \in P(R)\ \{0, 1\}$ determines whether the place or transition is shared
- $k : k \in N$ is number of customers (in terms of queuing systems, the size of population)

Therefore, our approach uses abstract descriptions to represent scenarios for the intended composite services. In addition to the informal approach for ASCT, we support the use of FPQSPN to describe a flow of a composite service in the composition architecture. This gives us a concrete mathematical model to analyze the performance of services running on the proposed architecture. Accordingly, ASCTs are transformed into basic Petri nets for verification of semantic correctness of the composed service. Once the ASCT is realized by selecting appropriate actual service instances, we extend this basic Petri net to FPQSPN based on their properties. A FPQSPN can efficiently depict a sub-diagram representing repeated processes by folding corresponding transitions into a non-shared one. Therefore, we can effectively evaluate the service scenario under consideration, using multiple instances of an identical service. As

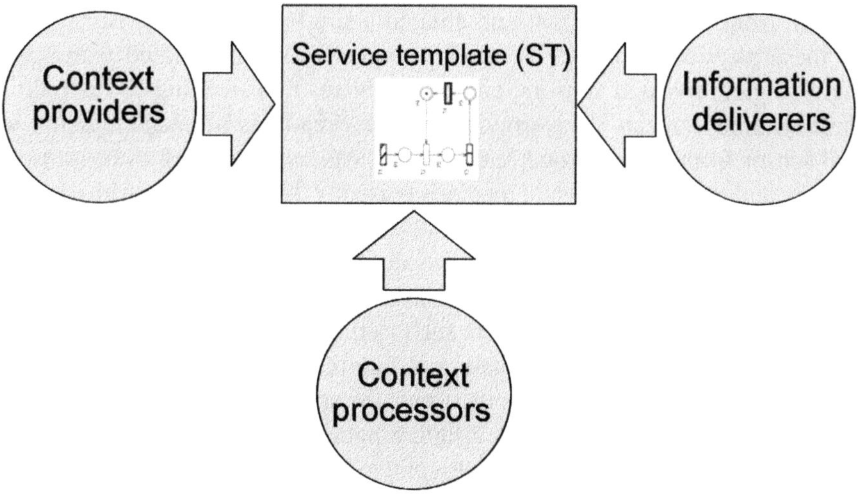

Figure 5. Pervasive service composition via ASCT.

discussed before, the three distinctive categories of basic pervasive services namely context providers, context processors, and information deliverers are the necessary components of an end-to-end composite service. Therefore, an ASCT in Figure 5 should involve at least three different abstract service elements, with each category contributing at least one element, in order to support an end-to-end pervasive service.

5. Performance Evaluation

For the purpose of performance evaluation, we consider an example based on the Smart Floor (Helal et al., 2005; Kaddourah et al., 2005). The Smart Floor service is a tagless indoor location system deployed in the Gator Tech Smart House. It provides unencumbered location tracking by employing low-cost pressure sensors deployed underneath the residential-grade raised floor to detect the user's current indoor location. As part of the package that offers assistance to mobility-impaired persons, at certain critical locations inside the house such as in front of the bathroom, the user's presence would trigger the smart house to open or close the door. To ensure that the smart floor detects the presence of the user in these critical areas, multiple redundant pressure sensors are deployed.

5.1 Enhancement in Reliability and Availability using Virtual Sensors

To evaluate how the virtual sensor framework improves reliability and availability, we simulated a deployment of eight pressure sensors in four adjacent tiles in front of the bathroom and created a singleton virtual sensor for each of these physical sensors. A basic virtual sensor was created using these eight singleton virtual sensors, and it aggregates their readings by outputting their arithmetic mean. The simulator fetches readings of each singleton virtual sensor from pre-recorded real-life data sets, within which every sensor is healthy and takes real pressure readings correctly throughout the entire logging session.

The purpose of the simulation is to evaluate the effectiveness of the compensation employed by the basic virtual sensor and observe how the VSQI reflects and is influenced by various factors such as relative error, number of singleton virtual sensor failures, and the pattern of failures. The simulator allows the injection of failures, either at specific times for specific singleton virtual sensor entities or according to predefined failure patterns, by using a random failure generator. Since at the time of data logging of the original data sets, all sensors were functioning properly; the induced failure allows us to evaluate how well the virtual sensors compensate for it.

We observed that the ability of a basic virtual sensor to compensate for failure is very effective. Our simulation shows that when one half (4) of the component singleton virtual sensors failed in a basic virtual sensor, the maximum relative error between the compensated output and the reference output, which is calculated using all available readings as if no failure occurs, is less than 1.50%.

The first set of simulation explores the correlation between VSQI and the relative error. As described earlier, VSQI is a measure of the quality of a virtual sensor, in terms of the probability that compensated values accurately reflect the readings should all sensors remain functional. Relative error is a different measurement of the quality of virtual sensor, which focuses on the deviation of the output from the expected value because of the internal compensation within a virtual sensor. Relative error can be heavily dependent on the specific data sets and the configuration of the virtual sensors. However, basic virtual sensors components deployed in realistic settings are usually spatially correlated, which leads to smaller and bounded relative errors. In Figure 6, we observe that in general, the VSQI and the relative error are inversely proportional to each other. As the relative error increases (usually due to more failures and heavier compensation), the VSQI decreases accordingly. From the plot, one also observes that when the relative error is approximately 2.3%, its corresponding VSQI value seems to be quite high. Upon inspection of the simulation logs, it was determined that this anomaly occurred due to temporary malfunctioning of one of the compensating sensors, which resulted in the output of spurious readings for a short duration. This incident is also reflected in the plot given by Figure 7, as discussed below.

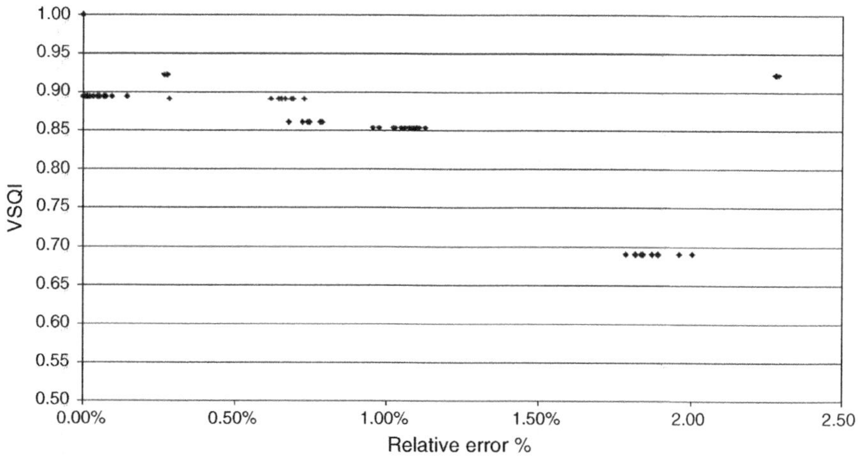

Figure 6. Relative error % vs VSQI.

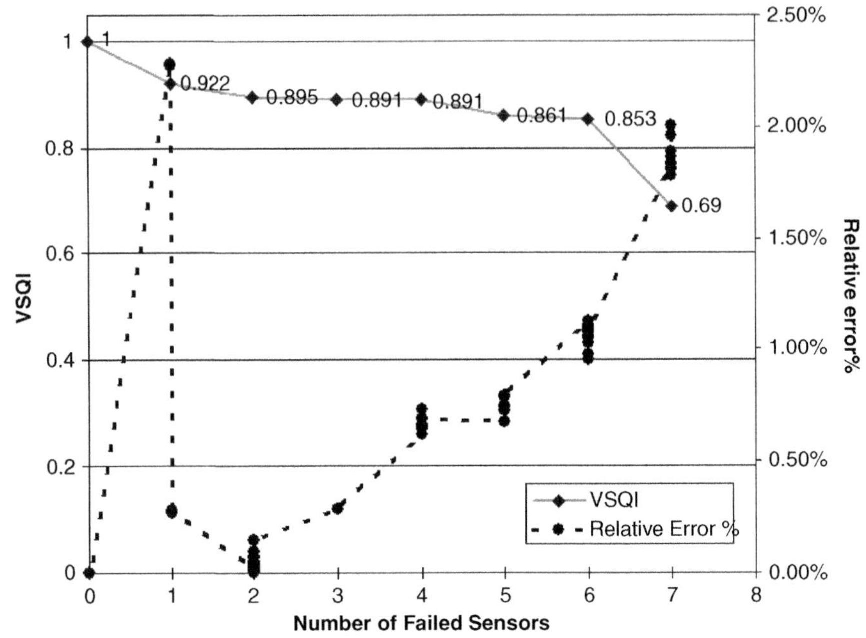

Figure 7. Number of sensor failures vs VSQI and relative error %.

Figure 7 demonstrates how the VSQI is affected by the number of failed singleton virtual sensors within a basic virtual sensor. The simulation shows that as more singleton virtual sensors fail, the VSQI decreases. Moreover, VSQI only decreases gently as sensors fail, which demonstrates the fault resiliency of the virtual sensor mechanism. Depending on the specific functioning sensor chosen to compensate for a failed one, the relative error may spike irregularly, but in general, the relative error increases as more sensors fail. One can also observe that there is a temporary increase in the relative error when the first sensor fails. Upon further inspection of the logs, it was determined that the compensating sensor experienced some temporary malfunction and as a result output spurious readings for a short duration, before resuming normal operation. As mentioned earlier, even under extreme widespread failures (87.5%), the relative error always remains bounded (less than 2.5%), which demonstrates the effectiveness of compensation using virtual sensors.

The results of the simulation also show the effect of failure patterns on VSQI. We introduce random singleton virtual sensor failures beginning at time 0, 180, 720, 2470, and 7920, but stipulate that all the sensors will fail within 80 epochs. The purpose is to explore whether allowing the sensors to accumulate a long correlation history improves VSQI when sensors start to fail. When the choice of the converging time constant a is kept low (we chose a=4), the weight of compensation plays a more dominating role. Figure 8 shows that the later the sensor failures are introduced, the sharper the drop in VSQI when sensors do fail. This phenomenon appears to be contradictory to our intuition at first, but upon further analysis, it is found that since the sensors that are used to compensate are not always placed at the exact same locality, the probability that two sensors exhibiting comparatively similar behavior always generate readings of equivalent magnitude decreases as the history accumulates over time.

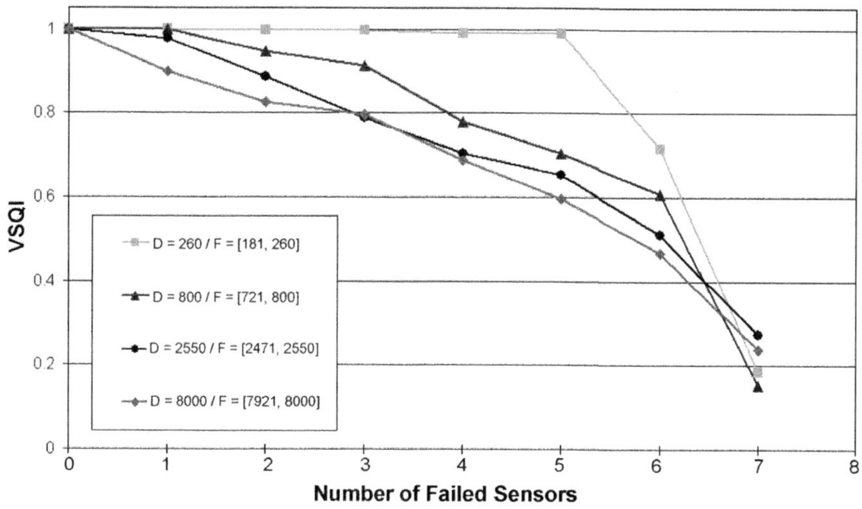

Figure 8. Effect of sensor failure timing on VSQI (D denotes total duration and F denotes the epochs between which the sensors fail).

5.2 Enhancement in Efficient Adaptability using WS-Pro with ASCT

To demonstrate how our WS-Pro/ASCT approach enhances the efficiency in adaptation, let us consider the following scenario. In an efficiency apartment within a smart apartment complex, the digital smoke detector senses an unusual amount of smoke in the air, which immediately triggers the fire alarm and initiates the sprinkler system as part of the fire emergency service. The fire started by an overloaded extension cord, however, quickly spreads along the cord and soon engulfs that part of the ceiling, where the smoke detector is located. As soon as the smoke detector is out of action, the smart house is no longer able to detect the fire and it shuts down the sprinkler system and it appears as if all hope is lost.

However, despite losing the most direct sensor for detecting a fire, the smart house is able to look up an alternative derived virtual sensor for fire phenomenon detection. This alternative derived virtual sensor includes an indoor temperature basic virtual sensor as well as a chemical basic virtual sensor. Fortunately, the indoor temperature sensor normally used for climate control picks up the unusually high room temperature, while the chemical sensor detects abnormal amounts of CO_2 inside the apartment. Thus, the alternative fire sensors are dynamically created and ready to be used for the end-to-end fire emergency management service. WS-Pro/ASCT that supports service re-planning and composition framework is able to quickly substitute a newly created context provider bound to the alternative fire sensor, for the original one associated with the destroyed smoke detector. Within a second the sprinkler is back on, and the automatic 911 call to fire department quickly prevents a tragedy in the making.

We next illustrate how our WS-Pro/ASCT approach models a service composition using ASCT and provides efficient service substitution.

5.2.1 ASCT for the Mission Critical Service.

In Figure 9, an ASCT for an emergency fire management service is associated with a set of actual service instances. Here, the service process for the ASCT consists of abstract service elements denoted with dashed filled boxes and flows denoted with dashed filled arrows. Once appropriate real services denoted with a box are selected by WS-Pro/ASCT, they are associated with an abstract service element. These associations are denoted using shaded arrows.

According to the scenario, the smoke detectors are quickly knocked out of action. The probe in WS-Pro/ASCT captures this exception and notifies the service composer to perform service re-planning. In this case, WS-Pro/ASCT does not need to create a new ASCT. Instead, based on the current service process of the ASCT shown in Figure 9, WS-Pro/ASCT searches for alternate actual service instances which are still alive and associates them with the abstract service elements. For example, the smoke detector is replaced with a derived virtual sensor composed of a temperature sensor and a chemical sensor. Similarly, the abstract service element "Responder" is re-associated with an automatic 911 caller. During this process, WS-Pro/ASCT considers performance data including reliability and availability.

5.2.2 Petri Net Model for Mission Critical Service in WS-Pro/ASCT.

In order to provide timely composition, the Petri net in our WS-Pro/ASCT approach is based on the FPQSPN model. Figure 10 shows the FPQSPN derived from the ASCT illustrated in Figure 9. The object with a double-line border represents a shared object such as virtual sensor, while the one with a single-line border depicts a non-shared object such as a unique software component.

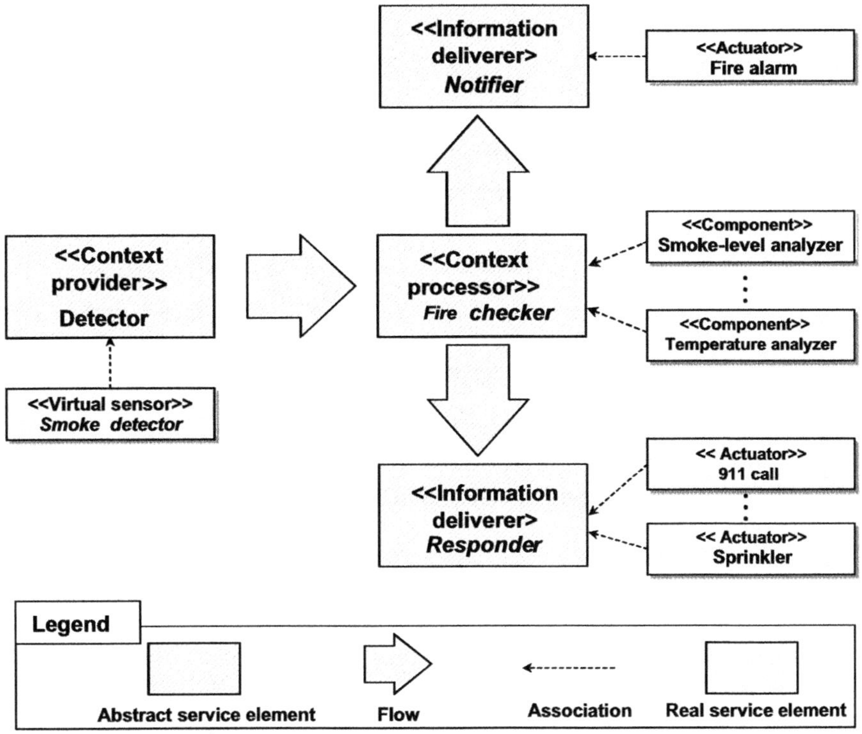


```
<sequence>
  <search name="hazard detector" attribute="fire"/>
  <copy from="virtual sensor type" to "detector type"/>
  <invoke name ="fire detector" operation="obtain environment data"/>
  <search name="hazard processor" attribute="fire" />
  <invoke name="fire checker" operation="identify fire" attribute="detector type" />
  <search name="emergency service" attribute="fire" />
  <invoke name="fire service" operation="actuator" />
<sequence/>
```

Figure 9. The ASCT for an emergency fire management service.

After a FPQSPN model is generated by transforming an ASCT, it can be used to measure the performance of a working mission critical service or evaluate the performance of a new service which was composed based on the ASCT. Note that the real data related to the t and tt tuples in the FPQSPN model are obtained from the actual service instances associated at present. Multiple instances of the context provider are presented as a non-shared object with the number of the service elements represented as k in the FPQSPN model. For this measurement for evaluation purposes, a simulator such as StpnPlay

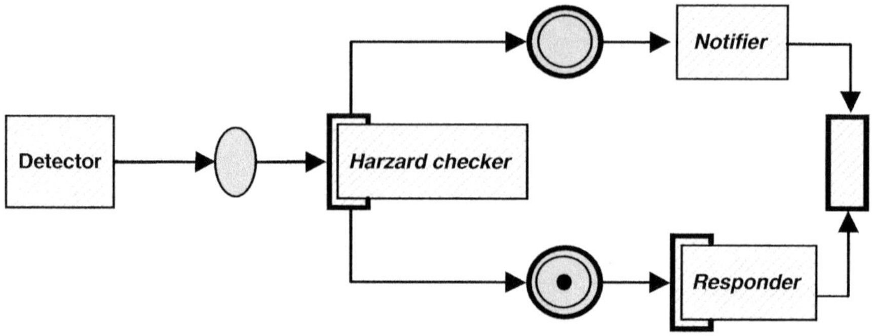

Figure 10. A FPQSPN for the ASCT in Figure 9.

(Capek, 2003) is used. However, we plan to have an integrated evaluation module of FPQSPN as part of the metrics computing module in WS-Pro/ASCT.

5.2.3 Effectiveness of WS-Pro/ASCT. We use simulation to evaluate the effectiveness of WS-Pro/ASCT in terms of adaptation. Without loss of generality, we assume that response times of all component services follow Poisson distribution. We compare two situations: with and without adaptation. There is a global deadline for a given composition graph. We consider the composite service to have failed when the global deadline is exceeded. When used without adaptation, component services execute on their own. With adaptation, we set deadlines for all the components services based on probabilities. For example, if the probability is 95%, then we set the deadlines such that services can deliver before their deadlines with a probability of 95%. When a component service exceeds its deadline, the adaptation mechanism kicks in and re-plans the execution. We compare the average failure rates against different probabilities, with and without adaptation, as depicted in Figure 11.

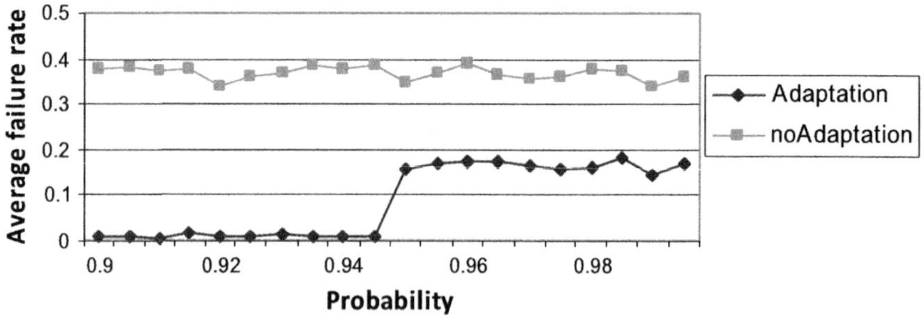

Figure 11. Performance evaluation of WS-Pro/ASCT.

From Figure 11 we can see that with adaptation, the service failure rate is consistently lower than without adaptation, which demonstrates that WS-Pro/ASCT is effective in enhancing service adaptation and improving its runtime performance. Note that the average failure rate increases with the probability. This is due to the fact that with higher probability we have less stringent deadlines. Hence, the execution sequence is longer before the adaptation mechanism kicks in. Moreover, the adaptation mechanism also takes some time to run. Therefore, the probability of missing the global deadline is much higher.

6. Putting It All Together: A Comprehensive Solution for Fault Resiliency

We present the overall system architecture in Figure 12 which provides a comprehensive solution for fault resiliency, by bringing virtual sensors and WS-Pro/ASCT together.

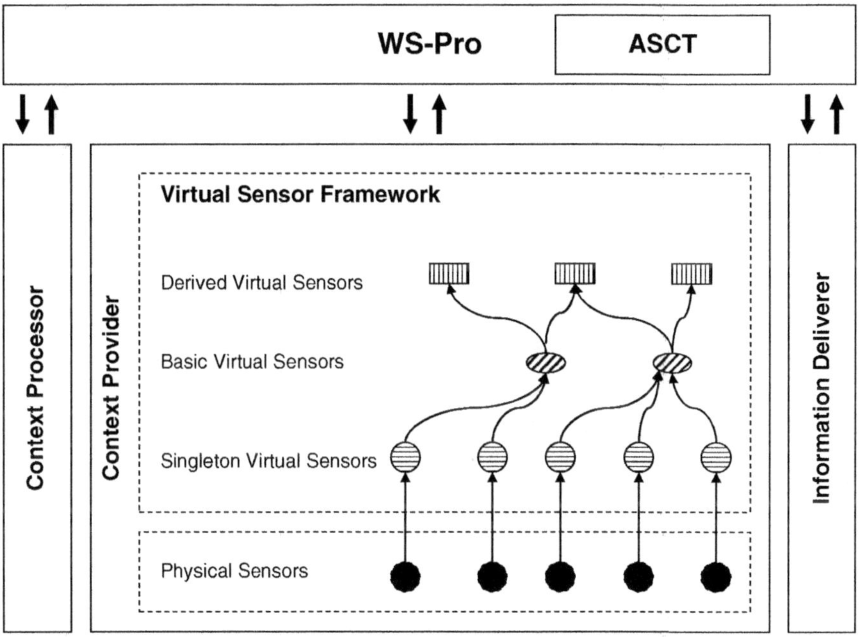

Figure 12. System architecture.

The inclusion of virtual sensors and WS-Pro/ASCT in a pervasive computing system allows it to continue functioning properly and degrading gracefully in face of sensor failures. Virtual sensors enable this by exploiting explicit redundancy (replicas) or indirect redundancy (correlated sensors) to compensate for the loss of data. However, when the system experiences extensive failures or becomes unstable, WS-Pro/ASCT kicks in and exploits redundancy at a higher level in the form of semantically equivalent services to stabilize the system. Even though both share the same objective of enhancing availability, each works at different levels and employs independent mechanisms, and their strengths complement each other. To provide a comprehensive solution to address the issue of fault resiliency, it is crucial to ensure a logical and smooth integration at various stages in the life cycle of the system.

The aim of the virtual sensor is to provide a robust and high-quality data source. The designers of a system look at the features of services in a system and decide which sensors to deploy, and the required level of fault resiliency as shown in Figure 13. The feature design dictates what kinds of basic virtual sensors need to be implemented, which include attributes such as the number, quality, and spatial distribution of singleton virtual sensors, as well as the aggregation algorithm to be used. This decomposition process gives a blueprint of which physical sensors to use, as well as where and how they should be deployed. On the other front, the functional requirements of services would justify the conceptual design and composition of multiple basic virtual sensors into a derived virtual sensor. Some of the reasons to design derived virtual sensors include dimensions which do not have means for direct measurement, the need for more comprehensive and abstract contextual information than raw readings, and the frequent reuse of certain aggregated data and information. As all virtual sensors are implemented as services, any of the singleton, basic, or derived virtual sensors can be a candidate in the service composition process. WS-Pro/ASCT service composition mechanism can match and choose these virtual sensor services based on various criteria, for instance, the fault-tolerance requirement, or whether raw data is preferred over comprehensive contexts.

These two pieces of the puzzle also work closely during runtime operation. As shown in Figure 14, each virtual sensor service by default constantly monitors its member virtual sensors and tries to compensate should any of them fail. Each virtual sensor also periodically self-evaluates its own quality, and all the associated derived virtual sensors constantly monitor the VSQI of their member basic and derived virtual sensors. As the result of our simulation shows, under realistic settings, virtual sensors are excellent at mitigating the effects of failures. However, in the case involving widespread sensor failures or malfunctioning of hard-to-compensate sensors, the quality of virtual sensor, as measured by the VSQI equations, might fall below certain predefined threshold $VSQI_T$. Should such a situation occur, the virtual sensor immediately notifies the WS-Pro module and requests for a service re-planning. WS-Pro works in tandem with the ASCT to re-plan the services utilizing the failed virtual sensor

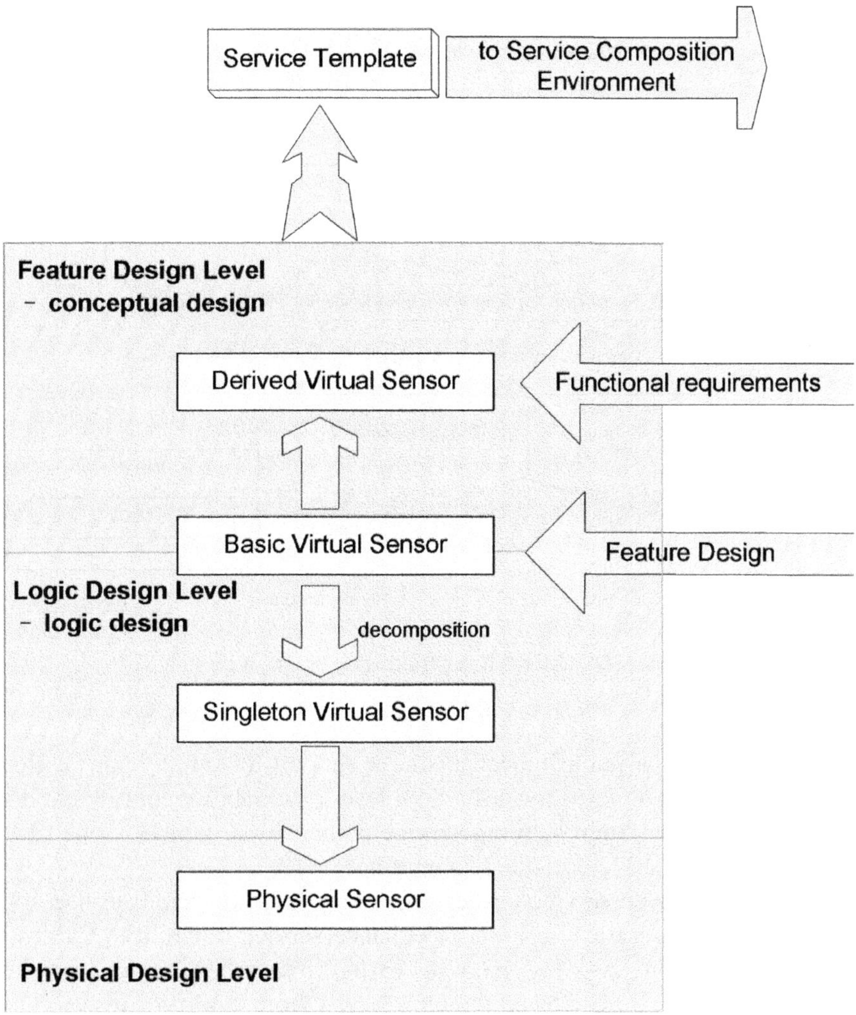

Figure 13. Design process for integrating virtual sensors into service templates.

and search for a replacement virtual sensor service to prevent interruption and breakdown of the overall service.

7. Related Work

Even though software reliability has been an active and critical research area for quite some time now, very limited research has been done to address the issue of reliability and availability of pervasive services in a SOA. Contraneo et al. (2003) have focused on changing dependability requirements and

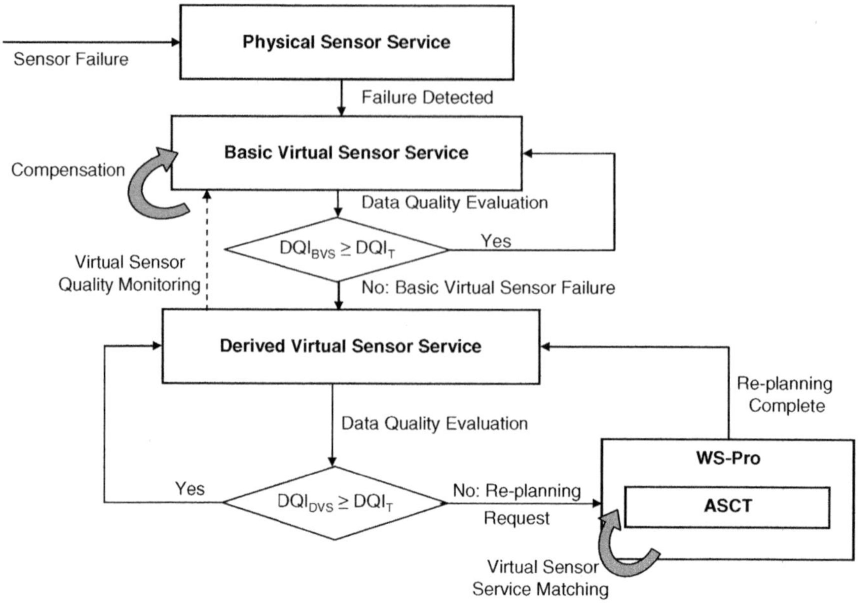

Figure 14. Monitoring performance of virtual sensors.

adopting a subscribe-publish mechanism for supporting fault tolerance. However, their work is targeted toward a high level of reliability control and does not address quality control during service composition, which is one of the core processes of service-oriented computing.

In the area of web services, researchers have tried to guide service composition using quality as a criteria to enhance service reliability. Cardoso's work (Cardoso et al., 2004) is one of the earliest researches in QoS-aware service composition. He borrowed the QoS model from workflow research and developed a static QoS-aware service composition environment. This model allows quantitative descriptions of non-functional aspects of workflow components from a service's perspective and computes the overall QoS of a workflow. However, in this methodology, the authors only considered static composition and did not address the more important and challenging problem of dynamic composition. Hence other related issues such as QoS monitoring and QoS-based adaptation are not supported.

Zeng et al. (2003); Zeng, et al. (2004) have presented a middleware platform which supports dynamic service composition. Their QoS model includes quality attributes which are different from the ones specified in Cardoso's model. The success of this model is based on the assumption that web service providers notify service requesters about their status. However, this mechanism is clearly inefficient in situations where network load is heavy, since notification messages can easily get delayed or lost. Zeng divided service composition planning into two categories–local planning at the task level and global planning at the execution path level. The architecture of this approach is very simple and the authors do not provide a concrete design of the trigger for re-planning. The authors also investigated performance of the service composition procedure but reported poor performance during re-planning.

The architecture proposed by Serhani et al. (2005) uses a two-phase verification technique performed by a third-party broker. This global broker publishes certificates attesting to quality of service providers and performs runtime measurement to compute QoS metrics as specified in the service interface. When the quality degrades, the broker notifies the service provider or initiates negotiation if the level of quality cannot be maintained. This architecture is still in a preliminary stage and the authors did not provide design details. Furthermore, this approach is not composition oriented so QoS model, service selection, and re-planning are not considered.

The broker-based architecture in Yu and Lin (2004, 2005) models the service composition problem as a Multiple Choice Knapsack Problem (MCKP). This methodology is different from other approaches as it takes the level of service into account during service selection. However, this approach is designed for the multimedia domain; hence, it includes some domain-specific attributes in the QoS model. Furthermore, this approach does not define a trigger for re-planning.

Several recent research papers have focused on virtual sensors and their role in abstracting physical data. Kabadayi et al. (2006) propose an intuitive virtual sensor mechanism for abstracting data from physical sensors. However, they neither provide an in-depth treatment of the possible types and functionalities of virtual sensors nor do they specify quantitative measures for monitoring quality of sensor data at runtime. Gu et al. (2004) present a formal and ontology-based context model using rules which describe context-aware behaviors. Hardy and Maroof (1999) give an architectural model for virtual sensor integration which provides a framework for designing and constructing virtual sensors using a layered architecture.

Alternate approaches for coping with sensor failure have also been proposed, such as Costantini and Susstrunk (2004), where sensor data is estimated by considering noise on the image that is captured by photo sensors, sensor fusion and failure detection (Long et al., 1999), and multi-sensor management and information fusion (Xiong and Svensson, 2002).

However, none of these approaches consider the utilization of virtual sensors for enhancing reliability and prolonging availability of basic pervasive services. Furthermore, there seems to be a distinct lack of availability of quantitative methods and formulae for measuring the quality of sensor services and ensuring that the data originating from them satisfy certain guarantees.

8. Conclusion

The introduction of service-oriented architecture (SOA) provided a clean and expressive interface that allowed easy application composition and dynamic self-integration. However, pervasive computing systems built on top of a SOA still suffered from failures and dynamic changes in the environment. In this chapter, we presented a comprehensive solution combining the concept of virtual sensors, which improved the availability and quality of data, with WS-Pro/ASCT, a performance-driven pervasive service composition framework, to enhance adaptability and fault resiliency. This integration created a systematic reliability control of end-to-end services by covering everything from software to hardware as well as from component to the entire system. It provided not only a reliable execution environment but also an analytical model to evaluate the impact of changes that occur. We also demonstrated the effectiveness of our solution through simulations using real-life data sets.

There are two primary goals that we are trying to achieve with our current effort and future work. We would like to accommodate more diverse and larger number of pervasive computing systems to be able to adopt our solution, so they can take advantage of the fault-resiliency mechanisms. We are also pursuing tighter integration between various pieces in our solutions, starting from a comprehensive design process all the way to the automatic, systematic integration of virtual sensors, service templates, and service composition framework at runtime. A representative list of work in progress is given below:

- More flexible and knowledgeable virtual sensors: We plan to construct sensor failure models that take into account factors such as deployment patterns of sensors and explore more efficient knowledge representation and management mechanisms.

- Standardization and systematic integration of WS-Pro/ASCT: We are currently establishing a standardized syntax for ASCT as a crucial extension to WS-BPEL (OASIS, 2007), a systematic transformation from ASCT to FPQSPN, and a tighter integration of FPQSPN into the WS-Pro framework. This will allow the streamlined automatic conversion of a business process to a service template and formal model.

- Better integrated software engineering practice for utilizing both virtual sensors and WS-Pro/ASCT: Templates in ASCT are often used to describe high-level end-to-end services, while virtual sensors are often defined based on the availability of physical sensors as well as functional requirements. There is a need for compatible meta-data that is flexible enough to accommodate both the needs to describe virtual sensor and service template matching. There is also a need for a systematic service decomposition method for identifying necessary service elements, critical services and group redundancy or crosscutting concerns. One of the procedures we are exploring is the Function-Class Decomposition with Aspects (FCD-A) (Chang and Kim, 2004), which allows software engineers to design a system with add-on crosscutting concerns such as availability and reliability.

- Integrated programming tools: Currently, the definition of various virtual sensors as well as abstract service templates is being constructed manually and separately. An integrated tool can embody the integrated software engineering practices mentioned above to improve the overall fault resiliency of the system created. Working in tandem with the automatic and systematic WS-BPEL/WS-Pro/ASCT conversions, the programming tools that we envision will allow designers to easily specify the business process and group sensors into virtual sensors, while the system automatically establishes the model, template, and dynamic service composition.

- Incorporation of self-learning virtual sensors: Virtual sensors have to be manually defined right now. Since the compensation and assessment of virtual sensors heavily utilize historical patterns and matching of probabilities, there exist excellent opportunities for employing distributed agents or rule-based reasoning engines to automatically define meaningful and useful virtual sensors.

References

Baresi, L. and Guinea, S. (2005). Towards Dynamic Monitoring of WS-BPEL Processes. In *Proceedings of the International Conference on Service-Oriented Computing*.

Bose, R., Helal, A., Sivakumar, V., and Lim, S. (2007). Virtual Sensors for Service Oriented Intelligent Environments. In *Proceedings of the 3rd IASTED International Conference on Advances in Computer Science and Technology*.

Bose, R., King, J., Pickles, S., Elzabadani, H., and Helal, A. (2006). Building Plug-and-Play Smart Homes Using the Atlas Platform. In *Proceedings of the 4th International Conference on Smart Homes and Health Telematics (ICOST2006)*.

Capek, J. (2001). *Petri Net Simulation of Non-Deterministic MAC Layers of Computer Communication Networks*. PhD thesis, Czech Technical University.

Capek, J. (2003). STPNPlay: A Stochastic Petri-Net Modeling and Simulation Tool. http://dce.felk.cvut.cz/capekj/StpnPlay/index.php.

Cardoso, J., Sheth, A., Miller, J., Arnold, J., and Kochut, K. (2004). Quality of Service for Workflows and Web Service Processes. *Journal of Web Semantics*, 1(3):281–308.

Chang, C. K. and Kim, T.-H. (2004). Distributed Systems Design Using Function-Class Decomposition with Aspects. In *Proceedings of the 10th IEEE International Workshop on Future Trends of Distributed Computing Systems*.

Controneo, D., Flora, C., and Russo, S. (2003). Improving Dependability of Service Oriented Architectures for Pervasive Computing. In *Proceedings of the 8th International Workshop on Object-Oriented Real-Time Dependable Systems (WORDS 2003)*.

Costantini, R. and Susstrunk, S. (2004). Virtual Sensor Design. In *Proceedings of Sensors and Camera Systems for Scientific, Industrial, and Digital Photography Applications V*.

Gu, T., Pung, H. K., and Zhang, D. Q. (2004). Towards an OSGi Based Infrastructure for Context-Aware Applications. *IEEE Pervasive Computing*, 3(4):66–74.

Hardy, N. and Maroof, A. A. (1999). ViSIAr: A Virtual Sensor Integration Architecture. *Robotica*, 17(6):635–647.

Helal, A., Mann, W., Elzabadani, H., King, J., Kaddourah, Y., and Jansen, E. (2005). Gator Tech Smart House: A Programmable Pervasive Space. *IEEE Computer*, pages 64–74.

Hummel, K. A. (2006). Enabling the Computer for the 21st Century to Cope with Real-World Conditions: Towards Fault-Tolerant Ubiquitous Computing. In *Proceedings of the IEEE International Conference on Pervasive Services*.

Kabadayi, S., Pridgen, A., and Julien, C. (2006). Virtual Sensors: Abstracting Data from Physical Sensors. In *Proceedings of the International Symposium on World of Wireless, Mobile and Multimedia Networks*.

Kaddourah, Y., King, J., and Helal, A. (2005). Post-Precision Tradeoffs in Unencumbered Floor-Based Indoor Location Tracking. In *Proceedings of the 3rd International Conference On Smart Homes and Health Telematics (ICOST2005)*.

King, J., Bose, R., Yang, H.-I., Pickles, S., and Helal, A. (2006). Atlas: A Service-Oriented Sensor Platform. In *Proceedings of the 1st IEEE International Workshop on Practical Issues in Building Sensor Network Applications*.

Long, T. W., Hanzevack, E. L., and Bynum, W. L. (1999). Sensor Fusion and Failure Detection Using Virtual Sensors. In *Proceedings of the American Control Conference*.

Ludwig, H. (2003). Web Services QoS: External SLAs and Internal Policies. In *Proceedings of the 4th International Conference on Web Information Systems Engineering Workshops*.

McCoy, D. W. and Natis, Y. V. (2003). Service-Oriented Architecture: Mainstream Straight Ahead. Technical report, Gartner Inc.

OASIS (2007). Web Services - BPEL Version 2.0.

Serhani, M., Dssouli, R., Hafid, R., and Sahraoui, H. (2005). A QoS Broker Based Architecture for Efficient Web Services Selection. In *Proceedings of the International Conference on Web Services (ICWS 2005)*.

Srivastava, B. and Koehler, J. (2003). Web Service Composition – Current Solutions and Open Problems. In *Proceedings of the International Conference on Automated Planning and Scheduling*.

Xia, J. (2006). QoS-Based Service Composition. In *Proceedings of the IEEE International Conference on Computer Software and Applications*.

Xia, J. and Chang, C. K. (2006). Performance-Driven Service Selection Using Stochastic CPN. In *Proceedings of the IEEE John Vincent Atanasoff International Symposium on Modern Computing*.

Xiong, N. and Svensson, P. (2002). Multi-Sensor Management for Information Fusion: Issues and Approaches. *Information Fusion*, 3(2): 163–186.

Yamato, Y., Tanaka, Y., and Sunaga, H. (2006). Context-Aware Ubiquitous Service Composition Technology. In *Proceedings of the IFIP International Conference on Research and Practical Issues of Enterprise Information Systems*.

Yu, T. and Lin, K. (2004). The Design of QoS Broker Algorithms for QoS-Capable Web Services. *International Journal of Web Services Research*, 1(4):33–50.

Yu, T. and Lin, K. (2005). Service Selection Algorithms for Web Services with End-to-End QoS Constraints. *Journal of Information Systems and e-Business Management*, 3(2):103–126.

Zeng, L., Benatallah, B., and Dumas, M. (2003). Quality Driven Web Services Composition. In *Proceedings of the 12th International Conference on World Wide Web (WWW 2003)*.

Zeng, L., Benatallah, B., Ngu, A. H. H., Dumas, M., Kalagnanam, J., and Chang, H. (2004). QoS-Aware Middleware for Web Services Composition. *IEEE Transactions on Software Engineering*, 30(5):311–327.

Chapter 10

INTRAVEIN – PARAMETRIC URBANISM

Brian Dale, Ioannis Orfanos, Pavlos Xanthopoulos, Gerard Joson
Design Research Lab, MArch Thesis Project, Architectural Association School of Architecture, London, UK.
{briancurtisdale, orfanos80, pavlos.xanthopoulos}@gmail.com, gjo507@mac.com

Abstract The chapter is about a form of networked urbanism distributed in east London, consisting of a system of infrastructural and leisure clusters of cellular units, combining as a connective tissue of bridges and islands, adapting, and negotiating as both a physical and an informational network. Embedded with self-learning behavioral and responsive systems, it allows for an intelligent choreography of soft programmatic spaces to create new leisure experiences, negotiating the changing effects of time, weather, programmatic, and crowd-dynamical inputs, extending parametric processes to drive urban performance.

Keywords: Distribution; Information; Space; Cells; Behaviors; Prototype; Networked; Parametric; Interface.

1. Introduction

Intravein is a proposal of a parametric urbanism for the region of Stratford, east London, UK, which explores the integration of adaptive spaces within a networked urban system while taking into account the dynamics of cultural, social, and economic flows. Carried out under the agenda of Parametric Urbanism at the Architectural Association's Design Research Lab, the research seeks to explore new forms of urbanism through the criteria of parametric design using Stratford and the surrounding 2012 Olympic development area as a case study. The research begins with exploring parametric design techniques in experimentation with the embodied flow of information within the city, leading to a proposal of an urban system consisting of infrastructural and leisure cellular units which combine into a network of bridges and urban islands, adapting, and negotiating both physical and informational networks. These cells are

W. Minker et al. (eds.), *Advanced Intelligent Environments*,
DOI 10.1007/978-0-387-76485-6_10,

embedded with self-learning behavioral and responsive systems, allowing for an intelligent choreography of soft programmatic space to create new leisure experiences. By negotiating the changing effects of time, weather, programmatic, and crowd-dynamical inputs, the parametric processes are extended to drive the urban performance (Figure 1).

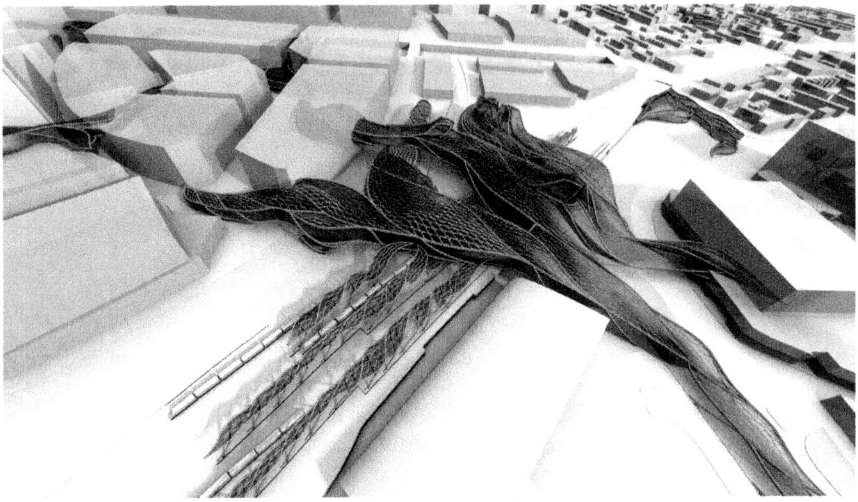

Figure 1. Aerial view of the proposed Stratford Bridge.

2. Description of Thesis Project

KNFRK propose a cellular form of networked urbanism, one where the large scale is made up of a series of differentiated elements distributed through the fabric of the city, where the urban parameters that feed the design are constantly indexed and used to drive the performance of the system. Through a neurological connection of these discrete elements the network gains the potential to constantly adapt to as well as itself adjust the dynamic conditions of the city.

This network consists of a system of infrastructural and leisure clusters of cellular units that combine as a connective tissue of bridges and islands negotiating the old and new Stratfords of east London as a physical and informational network. These clusters are embedded with self-learning behavioral and response systems, allowing for an intelligent choreography of soft programmatic spaces to create new leisure experiences that negotiate the index of changing effects of time, weather, programmatic, and crowd-dynamical inputs (Figure 2).

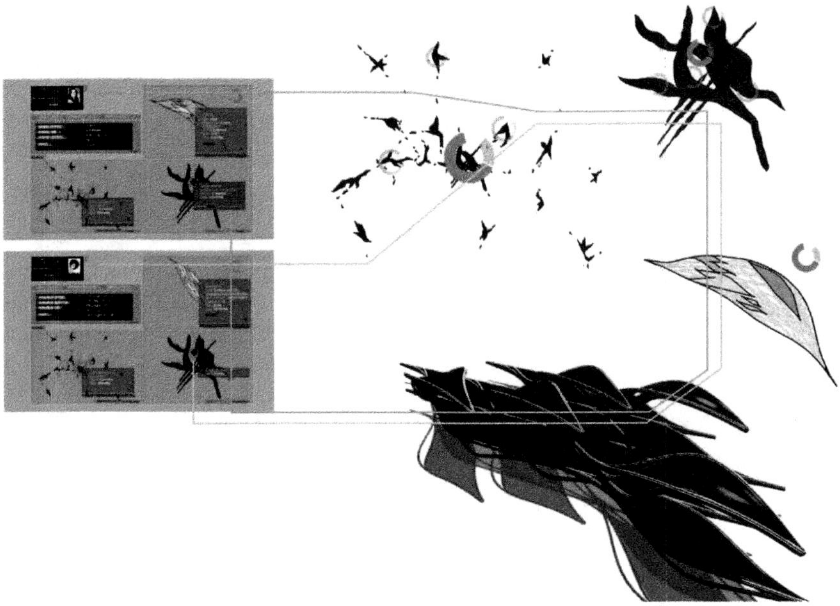

Figure 2. Diagram of the system of physical and informational networks.

3. Networked Behaviors

3.1 Intelligent Spaces

The system develops an intelligent infrastructure as a connective tissue to link the two poles of Stratford. Given the extensive existing and proposed commercial development in the site, a series of cells of leisure uses are proposed along a bridge and within islands in Stratford City. These cellular units are injected with certain behaviors, allowing one cell to adapt to different uses depending on the information that is indexed by the system, as a cinema reorganizes into a series of karaoke rooms. The project is scenario based, it cannot perform only in one way but rather must negotiate and reorganize according to the incoming information, taking on multi-state behavior.

The bridge at its most basic functions as a piece of urban infrastructure connecting two sides, but this typology is expanded to offer differing pathways, opportunities, and situations to the users as they inhabit the ever-adjusting bridgescape (Figure 3). This logic is then extended into the surrounding neighborhood fabric to develop a physically discontinuous but informationally linked archipelago of behavior spaces.

Thus, we have to be looking for a mechanism that reflects the criteria of the system. A mechanism of this kind can be considered a control mechanism of

Figure 3. View from the interior of the bridge space.

the entire procedure that should be built around a managerial intelligence that concerns the management of spaces during and after the design of parametric spaces. Is it possible to generate spaces based on algorithmic design and the logic of DNA so that the autonomy of forms and spaces would not be related with a predefined or even an autonomous system? On the other hand what is the potential of a relationship (exclusively or not) with a system that implies the active participation of users that interact and drive the system? This managerial intelligence could be found within the society itself instead of being left at the discretion of managers-creators.

In this hypothesis, a collective intelligence could reciprocate with the genetic material of architectural objects – abstract machines. "Once knowledge becomes the prime mover, an unknown social landscape unfolds before our eyes in which the rules of social interaction and the identities of the players are redefined" (Levy, 1997). In this social landscape humans participate in a dynamic, intelligence-based relationship with their environment (either artificial or natural). In a way we assume that management of machinic architectural objects could be realized by collectives. Of course this scheme could not guarantee a harmonic evolutionary model based on total control and guidance of a system. Rather the development of individual responsibility to navigate and communicate inside our broader social ecosystem is necessary, a "Fuzzy aggregate, a synthesis of disparate elements, is defined only by a degree of consistency that makes it possible to distinguish the disparate element constituting the aggregate" (Deleuze and Quattari, 1987).

3.2 System Networking

The proposal is structured around a system of networked elements with different scales and behaviors that are seeded and then grow within Stratford City as our case study. These elements concern activities of urban leisure such as entertainment, commerce, community activities, and public infrastructure. The main urban intervention is translated through input parameters to spatial output: evolution from the initial spatial unit of the single cell to the urban scale of Stratford (cell > cluster > island > urban network) while allowing for human-to-space interaction.

3.3 Networked Urbanism

Based on a fractalized logic of evolution, the system consists of cells as spatial units, of clusters as aggregation of cells, of islands as synthesis of clusters, and finally as an urban archipelago of networked islands (Figure 4). All these components are interconnected through specific scalar formal and informational relationships, but also through a shared negotiation of adaptivity and interaction, offering in this way a rhizomatic evolutionary perspective. Unpredictability becomes a basic factor of networked urbanism that parametrically enhances the urban experience and creates an ecosystem where cells, clusters, and islands interact without a linear dialogue with their environment and users.

4. Informational Experiments

4.1 Crowd Dynamics

Crowd dynamics are used as a simulation technique in order to explore organizational behaviors that evolve in time. The generation of organizational models is based on various parameters (number, cohesion, separation, alignment of direction and speed, type and number of targets) that take specific values according to time (day/week/season). Time and space criteria are merged as time-based parameters feed the behavior of agents through attraction points. The attributes of these targets are differentiated following various desire properties that are then translated to local spatial conditions and crowd desires.

These studies suggest that optimum organization of space varies in time and that parametric tools can simulate and reorganize qualitative–quantitative spatial and temporal characteristics.

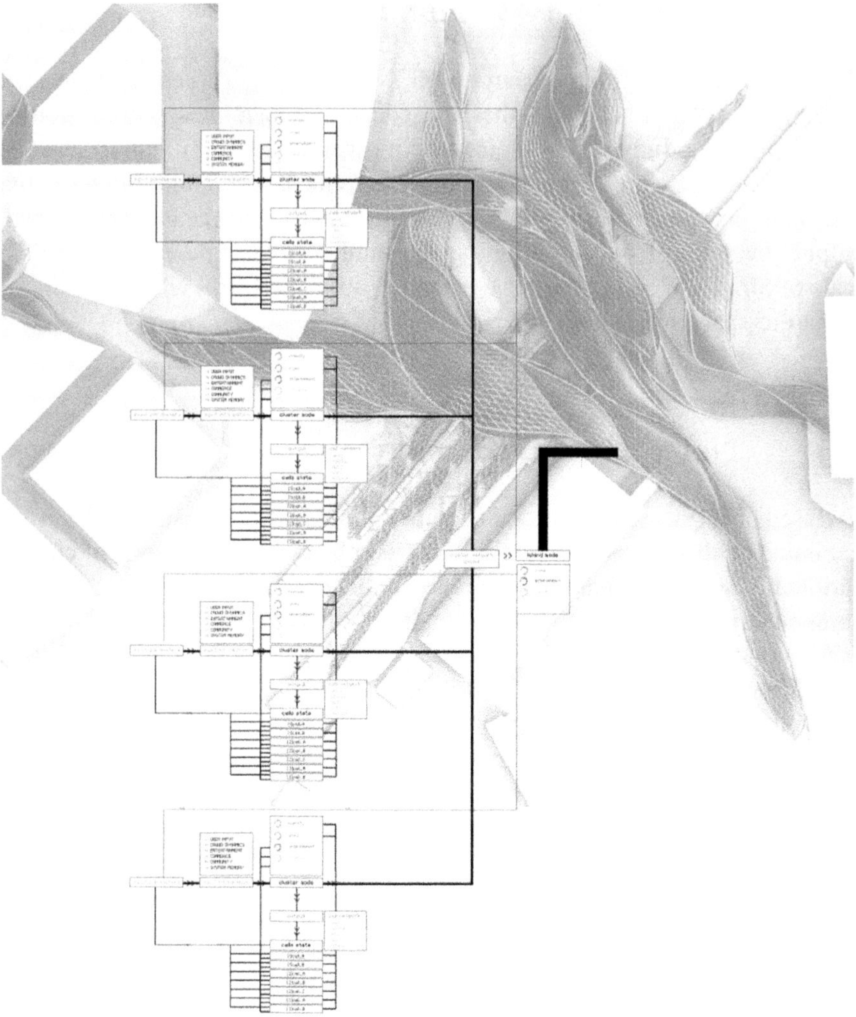

Figure 4. Diagrammatic structure of the information flow between spaces: synthesis of cells, clusters, and islands.

4.2 Human Interface

Interfaces play an important role within the scheme. It is about a man–machine interaction whose center is activated inside the system, a mediator linkage. Through my choices inside an informational structure made using an interface, I participate in a collectivity that altogether informs a system with data. This data affects not only the performance of a space but primarily the algorithm code that controls the space. So being aware of my input, I contribute

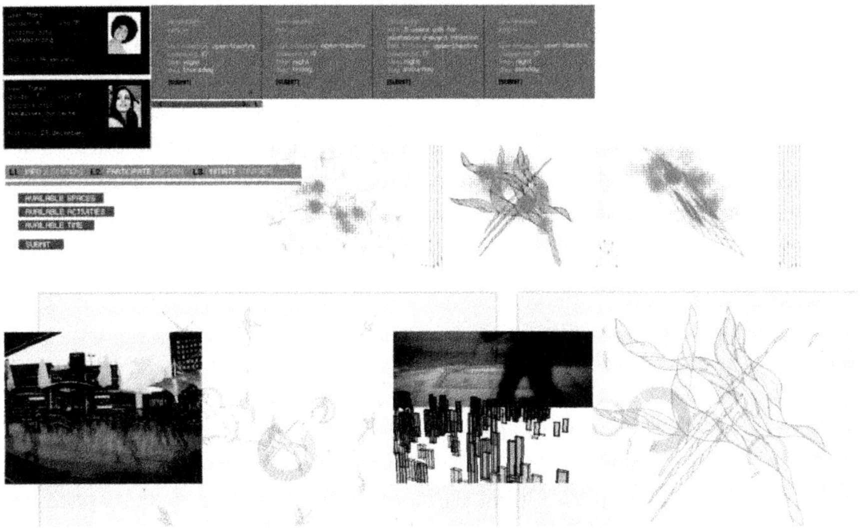

Figure 5. Snapshots from human interface and informational experiments.

to an intelligent collectivity that deals with the system as a collective input (Figure 5).

Intelligent collectives, "similar to vernacular archetypes and prototypes that have been developed and adapted for different sites, environments and individual requirements" (Frazer, 1995), can interact in the complex context of contemporary urban life. In this way a decentralization of the system follows, implying loss of control by the author. In reality, it "removes the architect from the design process: giving the environment the ability to design itself and to have autogenetic existence . . . is about a new kind of architecture without architects (and even without surrogate architects)" (Negroponte, 1975). The triptych intelligent collectivity-reorganized space interface can be reduced to the primal triptych human–machine interface.

The idea of an interface as part of an urban network becomes important, constantly feeding the parametric behavior of the system with data. The digital extension is an integral part of urban activity and merges intensively with spatial urban experience. The code of the system becomes open, culling feedback from people's desires. This could be translated to become conscious of taking responsibility about our social relations parallel to deep embodiment of technology. A parametric space managed by collective intelligence implies a complex system and an abstract communication relationship between user and interface. "As clusters of tools, procedures, and metaphors, technologies configure a platform for discourse and ideology. Such a technical-discursive ensemble is modifiable through politics, yet it has political orientations built into it system" (Crandall, 2005). The potential of all of these references is related to a collective organizational logic that accelerates social connections. Offering open procedures to users is not a risky decision, but on the contrary it is necessary to deal with specific existing conditions and to create new ones as well. "The architecture of the exodus will give rise to a nomadic cosmos that travels the universe of expanding signs; it will bring about endless metamorphoses of bodies; Far from engendering a theater of representation, the architecture of the future will assemble rafts of icons to help us cross the seas of chaos" (Levy, 1997).

4.3 User Input Indexing

Consumer becomes producer in an automated system.[1]

The connection between MAX/MSP and Flash through a user interface allows for the adjustment of global parameters concerning physical space from the users ubiquitously and at the same time enhances urban experience as an information visualization medium.

Personal involvement inside a networked urban field functions similarly to Internet online communities that share desires, but in this case they share desire about their space, their activities, and therefore their actual urban experience and interaction. This attachment emerges between user and system that exchange information by sharing the strata of space and time.

The user interface offers the navigation through information that deals with space, activities, and time. This structure allows human interaction to have an impact on physical experience as user interface input is connected with the adjustment of physical space.

The access to information is categorized into three levels of navigation and interaction. First, the user receives information through location-based navigation. Second, the user participates in the system by inputting his or her desires. Third, the user initiates the response of system behavior by trig-gering new activities.

4.4 Information Distribution

Inside a complex system of networking different locations, scales, and times, the organization of information is the first crucial subject. In order to connect information with space the basic categories become the foundation of the system. It is not about reducing complexity, but about simplifying the understanding of complexity through setting fundamental driving forces. The selected information categories mix space parameters such as crowd behavior, activity intensity, or programmatic uses (entertainment, commerce, community) with relevant information parameters.

Urban, island, and cluster choreography of information distribution are based on the above information categories and are applied as group of parameters of specific locators and at the same time as group of parameters of the relationships between the locators themselves (Figure 6).

4.5 Programmatic Parameters

The parametrization of program is realized through the codification of spatial programmatic uses with selected main parameters. Area of space, motion of users inside that space, sound in that space, use in that space, time factors of that space, and the potential of interacting with an interface become distinct as main parameters. The next step is the analysis to subparameters in order to proceed to measurable spatial qualities. Area concerns the area size, its shape and proportions, and the open, semi-open, or closed state of the space. Motion refers to the speed of movement through space and directionality of the crowd. Sound concerns frequency, loudness, duration, and rhythm of the environment. Use is categorized into entertainment, commerce, community sport, and infrastructure. Time is analyzed as the duration and time of day of the activity in the space and the time of responsiveness that is needed. Interface involvement is about the adjustability of that space and the potential of offering information to users through an interface (Figure 7).

4.6 Fluid Movements and Pulses

Our visual indexing records the past events in memory, but the current view is set to the present moment. By implementing a motion tracking system we are able to translate present movement into visualization with an increased lifespan. As the camera records the motion within the frame, zones of change give birth to sets of particles in the system who are encoded with behaviors of growth, atrophy, and a finite lifespan. The reactive system allows traces of recent events to be left behind, extending the moment of the past present.

The complex flows of bodies through a city carry embedded information within their patterns of movement. Fluid algorithms and particle physics are used here to simulate the dynamic conditions of the crowd. Seeing crowds as fluids allows for us to conduct digital research into the interactions between certain user types, focusing on the laminar and turbulent flows generated by the system and on the crowd as a whole, adding a layer of understanding to the neighborhood conditions of the agent-based crowd-dynamic simulations.

Parametric adjustment of the conditions of the simulation allows for time-variant conditions to be taken into account and the results studied relative to individual output. Here trajectories and points of attraction set up the simulation where fluid bodies interact according to the implanted values.

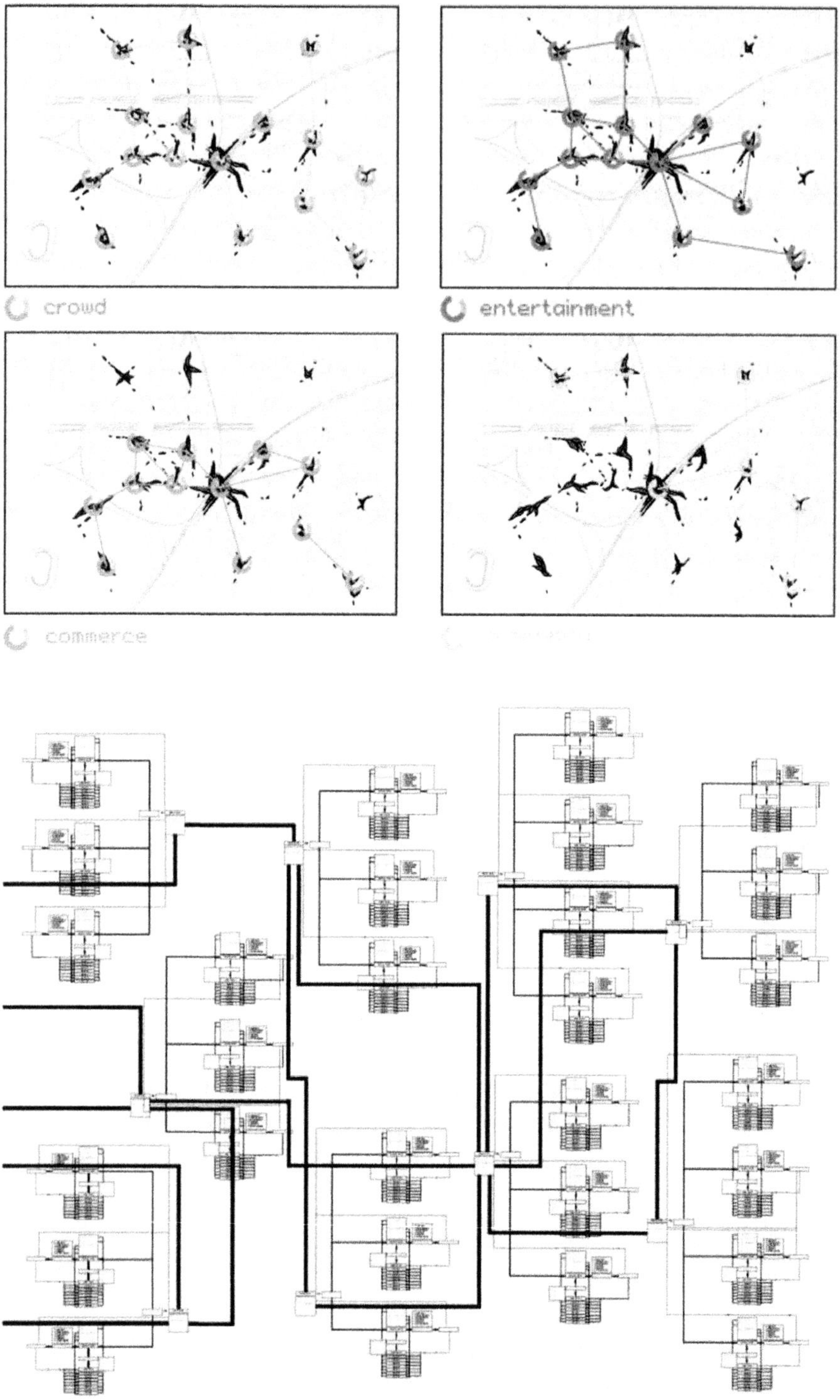

Figure 6. Urban relationships according to parameters (crowd, entertainment, commerce, community) and Information distribution through the network of clusters to islands and finally to the urban archipelago as a whole.

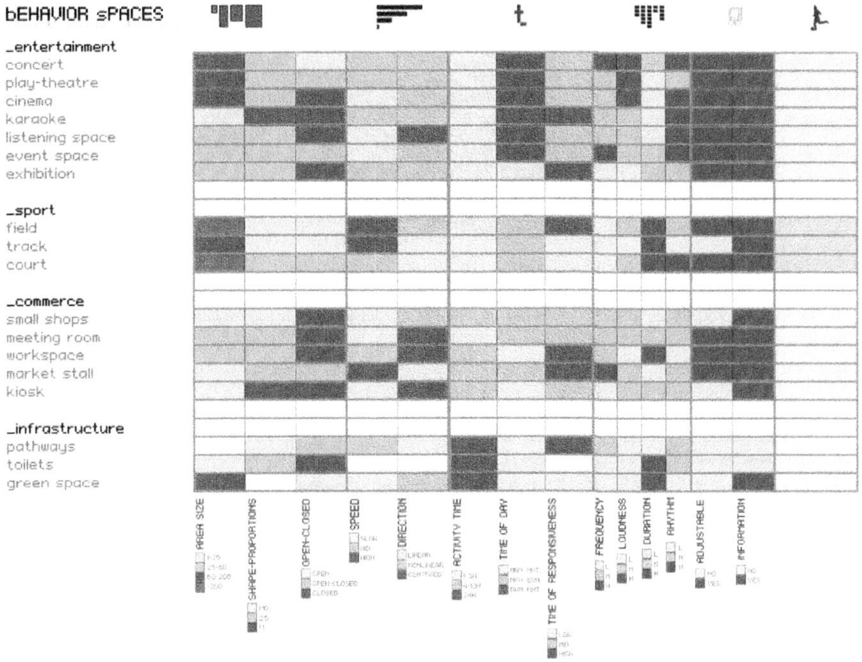

Figure 7. Programmatic parameters in relation to space behaviors.

4.7 Density Driven

Fluid algorithms provide a system in which local change is reflected through the global whole. Each voxel [volume pixel] that makes up the digital fluid reads the parameters of its neighbors and adjusts its condition accordingly, generating gradient flows across the fluid container. These fluid motions and their adjustment to local conditions provide a way to translate characteristics of crowds into a system that allows for the simulation to be adapted to drive other parameters of the design. Here the density values of the voxels are connected to corresponding cells with a discrete range of possible motion (Figure 8). Higher density values correspond to higher crowd densities and cause the cells to open in response, revealing the activation, speed, and trajectories of the users.

5. Space (in) Formation

5.1 Negotiating Behaviors

Multi-state behaviors emerge from a system in constant negotiation, adjusting between present and future conditions while progressing away from

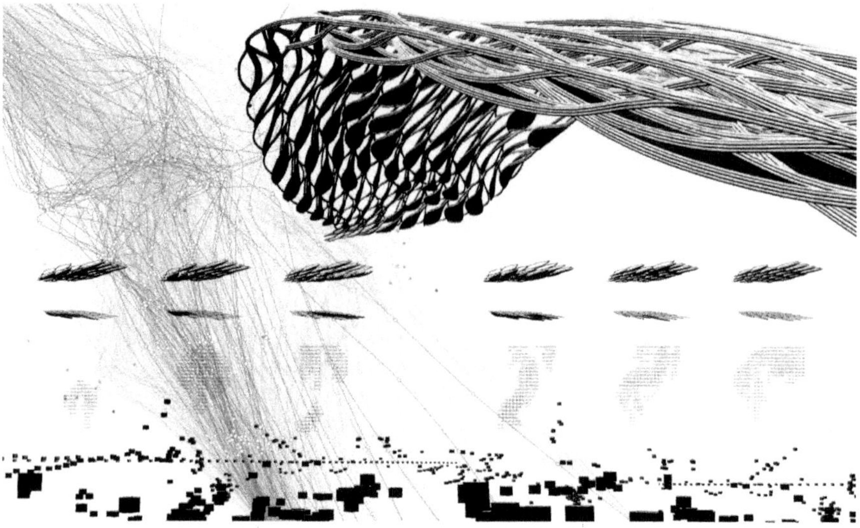

Figure 8. Space formations according to variation of fluid algorithms.

previous states. This dynamic equilibrium allows for low-level changes in the system to have global effects, maintaining a partial memory of past events and allowing for a conversation to occur between the states.

5.2 Splinal Logic

The splinal logic of constructing and merging splines of beams together was developed by creating specific properties that a cellular formation should follow. Bundling is the integral method of creating the infrastructure of the beam, where depth through accumulation is achieved. When a beam needs to split into different directions, the diverting property of the splines will allow the bundles of the beam to follow a specific trajectory. Because of the bundling ability, even if a beam splits into two parts, structural integrity can still be retained due to the redundant density of splines that the beam started with. Twisting in the system was used to create specific directional flexibility on the bundles created. With the twisting property, a vertical bundle can transition to a horizontal bundle that would allow the beam to flex and move in a different way. Depending on structural requirements, a spline can blend into other splines when a lesser amount of support is needed on a certain area; thus, material optimization is achieved.

5.3 Cellular System

A more coherent cellular element was derived based on the splinal logic studies. Following a straight trajectory of lines, the 1CELL unit is formed, its horizontal elements merging with diagonal splines enclosing a space. With the directionality of the cellular design, a simple rule of addition was followed to produce the next three sets of cells. Starting with a single cell, a second one is added at a 45° angle on any of its four edges so as to merge the two by sharing similar splines. With the addition of a third and fourth cell, the 3CELL and 4CELL can be derived, respectively. To create transitional spaces from enclosed to open both the upper and lower splines of the enclosed cell are merged with the lower splines of an open cellular space (Figure 9).

With the use of singular units, deformation can occur in a more local scale that ripples through neighboring elements. The flexibility of the system is seen with the ability of the cells to nest within larger cells, providing for a system of adjustability and easy deployability. Given a specific large cell to fill, the four types of cells can be easily paired up to give the best possible formation for the given space.

5.4 Parametric Relationships

Catia software was used to catalog the static and kinetic programmatic cells that then are applied with localized programmatic criteria to form clusters nested into infrastructural cells. The semi-enclosed interior of the resulting space is defined by the placement of floor and canopy cells. Each cell type follows specific rules controlled by relationships that allow for a range of adjustment to fit the cells to specific formations. The cells stay within a defined range of possible values of dimensional parameters, allowing them to share similar geometries but provide differentiated spatial outputs.

5.5 Cells

There are four main types of programmatic cells:

- single programmatic cell: a simple structure that is able to define small semi-enclosed spaces within the larger array, it is embedded with kinetic behaviors that allow it to adjust the height and position of its structure to accommodate various configurations;
- double programmatic cell: defined as two subtypes, one semi-enclosed and able to adjust by vertical motion alone, the second begins completely enclosed but is able to open to surrounding spaces with a combination of vertical movement and rotation;

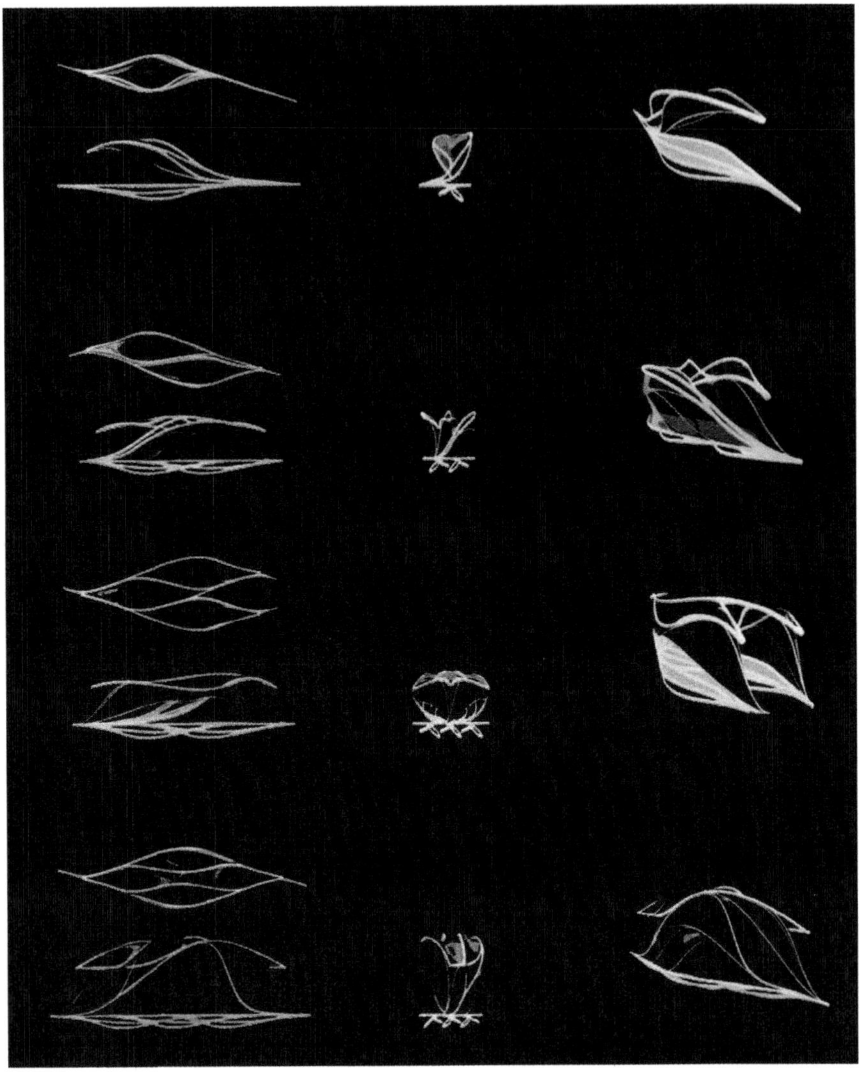

Figure 9. Typology of cells: 1cell, 2cell, 3cell, 4cell.

- triple programmatic cell: also developed into two subtypes, the first semi-enclosed and able to adjust the upper portion of the cell, while the second is able to perform as a walkable ramped connection to an upper level or as component in a varied landscape;

- quattro programmatic cell: entirely enclosed, these larger cells remain static in their structure to allow for more spatially demanding uses to coexist in the clustering formations.

5.6 Canopy Cells

Where the programmatic cells' states are determined by the cluster to cluster relationships within the network, the canopy cells respond to their immediate conditions providing a less hierarchical system of response to changing conditions. By reading the states of their neighbors as well as the movements and requirements of the fluctuating crowds, single cells are able to adjust their position accordingly, constantly adjusting the conditions and experience of the space below. These variations transfer across the global level of the system by passing information at the local level, resulting in an ability to adjust ambient spatial conditions through the resulting patterns of porosity.

The catalog of static and kinetic programmatic cells is applied according to localized criteria to form clusters that act as nodes within the system. The individual cells within these clusters respond to the positioning of their neighbors, allowing the cluster to take on varied formations. These clusters are then nested into the larger infrastructural cells, with the semi-enclosed interior of the resulting space defined by the placement of floor and porosity of canopy cells.

5.7 Robotic Prototypes

A series of manual and robotic prototypes were developed in order to explore the possible configurations of the kinetic cells in parallel with inverse kinetic digital animations. The digital and physical prototyping processes were explored such that the results were fed back into the corresponding development of the next prototype version. Initial prototypes attempted to replicate the initial behavior and range of movement of the animated typical programmatic double cell. In order to achieve this motion a simple linear pull mechanism was implemented to control the position of the structural arms of the cell, and with a short adjustment a far greater change was achieved. Adjusting the balance between stiffness and suppleness of the members allowed for the prototypes to be fine-tuned to desired behaviors (Figure 11).

The series of prototypes was very crucial for the development of the research and particularly is the means of moving from the state of an experimental conceptual proposal to the level of realization. The emergence of prototypes that reveal kinetic, interactive, and structurally efficient qualities is a crucial step to explore methods and techniques that in direct future and under specializing studies could lead to the technology required to realize such spaces. The experimental prototypes bring the conclusion that from the architectural point of view we are not far at all from building such an intelligent infrastructure.

5.8 Prototypes: 2CELLv.1–2CELLv.2

This first manual prototype 2CELLv.1 brought about some interesting developments in the design of the kinetic cells as the performance of the model was now related to the properties of its material. Through this material computation a twisting motion was added to the repertoire of kneeling/opening initially proposed through digital animation. The states of the cells are linked to the index of incoming information within the network. As the clusters communicate the relevant data of time, crowds, desires, program, and weather they determine the current state of the individual cells.

The supple robotic prototype 2CELLv.2 developed as a way to link the digital indexing of data to the physical manifestation of the cell state. Attraction points within the network are adjusted to the varying activation scenarios, and the localized conditions are then fed to an Arduino micro-controller via an interface developed in Max/MSP (Figure 10). Density within the digital diagram is translated to the position values of the servo motors that take over control of the linear pull, allowing for precise control and relationship of the pair of double cells. This state data is then fed back into the system through communication with the other clusters within the network and at an interface level to distant users within the urban network (Figure 11).

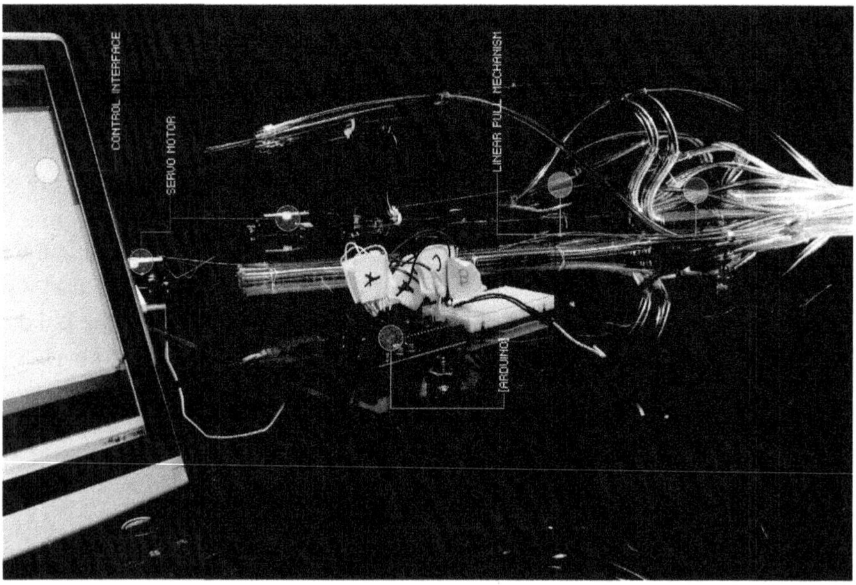

Figure 10. Robotic prototype 2CELLv.2 actuated with an Arduino micro-controller via an interface developed in Max/MSP.

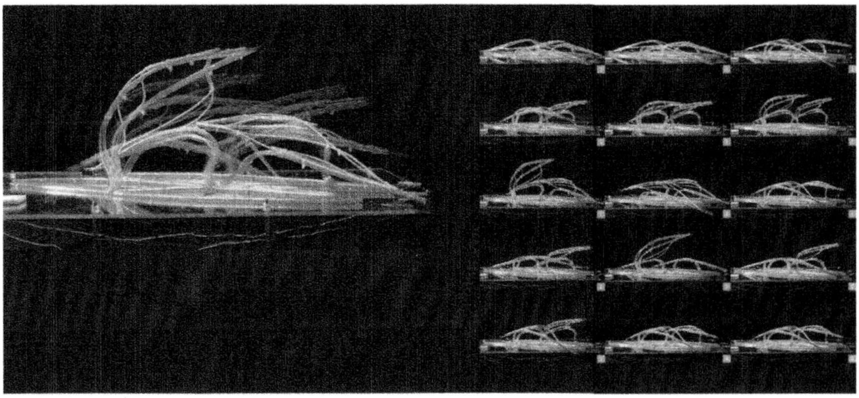

Figure 11. Spacial variations of prototype 2CELLv.2.

5.9 Prototype Canopy v.1

This prototype explored the possible configurations of a single canopy cell and its relation to the overall organization of the prototype array. Using similar materials as the previous cellular prototypes, the motions of the canopy cell are explored in a physical manner in order to bring out certain tendencies in the system to inform further development of the cellular type. Each cell is able to act independently, but the position taken is relative to the neighboring cells, working with the same principles that were used to initially study the fluid movement of the canopy array. These variations transfer across the global level of the system by passing information at the local level.

6. Distributed Responsive Leisure

6.1 Stratford

East London is preparing for an explosion of development in coming years, the time frame of which has been accelerated by London's successful bid to host the 2012 Olympic Games, whose site sits adjacent to the proposal.

Stratford Town Center is a major transportation hub for east London, currently providing connection between northeast England and London, with trains, buses, and the underground converging at Stratford Station and London City Airport a short DLR ride away. This is about to be expanded as the new International Train Station is constructed, extending Stratford's reach directly to mainland Europe.

6.2 Urban Network of Negotiation and Leisure

The addition of an urban leisure park at the point of connection between the two Stratfords would augment the existing and proposed development, offering an infrastructure that adapts intelligently to constantly changing scenarios. This intelligent infrastructure is set up as a parametric urban filter that creates unpredictable connections and follows economic, social, and cultural dynamics. A bridge is the continuously responding structure to these changing trends. It adapts, creates space, and even takes part in the negotiation between the two Stratfords. Leisure activities are proposed as an urban enhancement of the centralized commerce poles of old and new Stratford.

The result of this parametric urbanism is a network that spreads through old and new Stratford. Instead of a centralized mega intervention, a flexible urban network is articulated around the central station of Stratford that connects the two local poles. Urban islands are proposed in strategically important locations of the area, following the logic of an urban networking system that is realized as both a circulatory and a digital network (Figure 12).

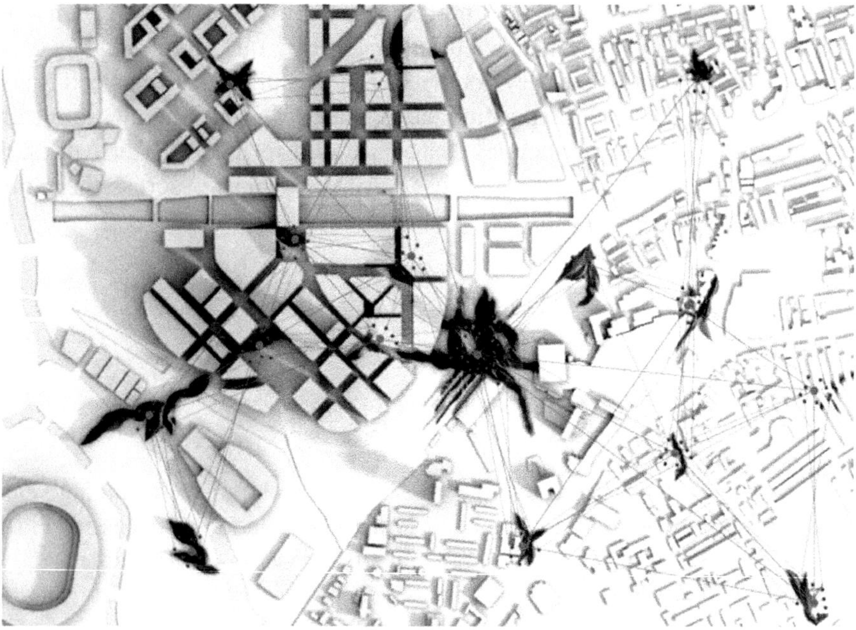

Figure 12. Urban and information network in Stratford, London.

6.3 Circulation Network

The proposed urban network of Stratford is first of all a circulation network. Its islands are located in proximity to each other in order to allow the movement between them. The old area of Stratford (east), the main bridge area, the new Stratford City (west), and the new Olympic site are connected in a common urban environment. The logic of dispersed urban intervention enhances the in-between places as well and circulation becomes vital, allowing a degree of freedom to users in contradiction to a centralized urban masterplan.

In addition, the placement of the network's nodes takes into consideration the existing and future main circulation directions, creating a "bridge" between existing space experience and the proposed adaptive urbanism.

6.4 Digital Network

On the other hand, the digital network that is developed concerns more translocality than locality, the way that physical world is connected to the system in terms of information. Therefore, the urban islands are interconnected and the flow of information between them is based on specific information categories derived from crowd behavior, entertainment, commerce, and community activities that translate the existing urban activities to the network's nodes. The digital network apart from the informational characterization of the urban islands and the information distributed between them extends also to ubiquitous human interfaces that provide to users access and interaction with the system's networking.

"Interaction, between the users and the pieces of information they handle, may become information itself. It is the automatic translation of these interactions that is taken over in the fuzzy interfaces. Accordingly, it is not only the interface graphic design that has to be conceived, but also the design of the structure of information itself" (RU3, 2006).

6.5 Localized Programmatic Hierarchy

The connection between global information parameters with the various areas involves the specific local conditions that differentiate each zone among the network. The daily urban experience within the network is not a homogeneous one. Therefore the formation of the islands takes into consideration the local conditions and becomes the extension and the augmentation of the existing urban environment. The islands of the proposal are placed in old Stratford (Great Eastern rd, Broadway triangle, High St., West Ham lane, Public Gardens, Vernon rd), in new Stratford (South Station square, Stratford City park, Retail square, Central spine, Olympic village, West Leyton public space, toward and by Olympic stadium), and in the bridge that connects both Stratfords (Stratford Bridge). An island is also placed on Maryland St. in the neighboring area of Leyton, suggesting the potential of a further expansion of the network into greater London.

Each island feeds and interfaces differently, with the digital networks producing its own localized programmatic hierarchy of entertainment, commerce, and community activities (Figure 13).

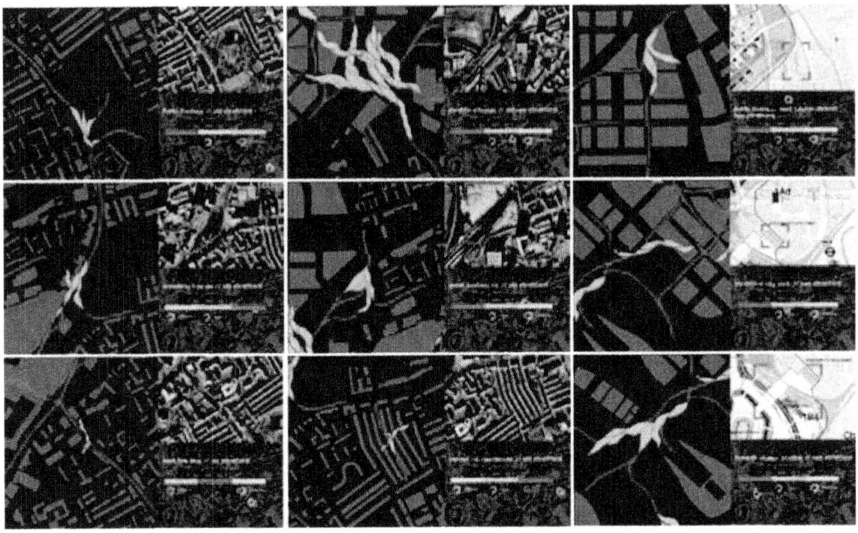

Figure 13. Differentiated programmatic hierarchy of areas of the proposed network.

6.6 Stratford Bridge

The Stratford Bridge island is the interface of social, economic, and cultural forces of old and new Stratford. Additionally, as an architectural proposal it combines all the spatial qualities that could be faced across the urban network (Figure 14). Apart from being a physical connection that facilitates the movement of vast amounts of people, it is a leisure park that hosts entertainment, soft commerce, and community activities in a larger scale than the other islands of the proposed network. It also extends organically to the proximal bus and train stations (Figure 15). The programmatic cells are the spatial units that cluster together to host human scale activities. The clusters are then attached to larger infrastructural bridge cells, the synthesis of which forms the urban island.

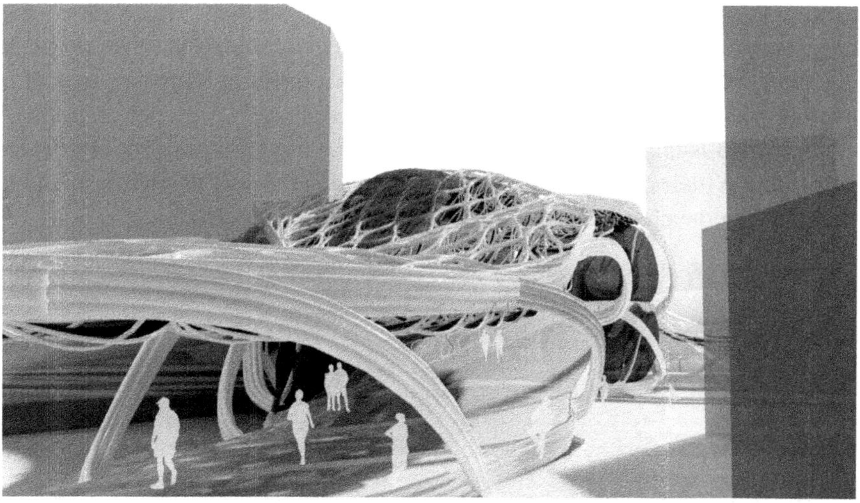

Figure 14. View of the entrance to Stratford Bridge.

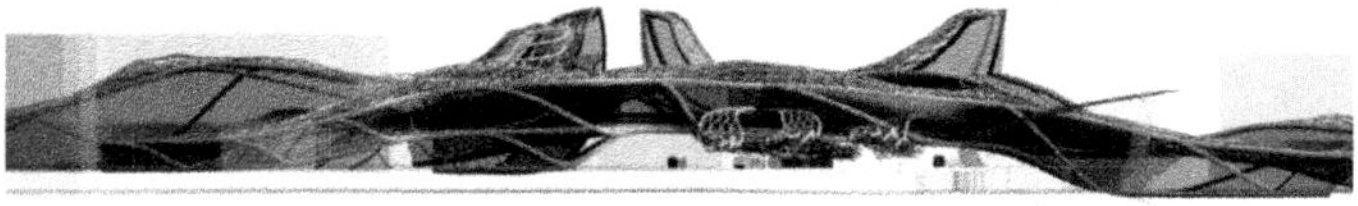

Figure 15. Elevational view of the bridge that connects new with old Stratford.

6.7 Cluster Networking

The information distribution is gradated to different spatial scales. Each island is itself another network of interconnected clusters of cells. The clusters are linked to each other according to basic information categories (intensity of crowd activity, crowd behavior, programmatic activities) based on interrelations that take into account local and spatial criteria. Proximity, location, programmatic tendencies, scale, and proportions are properties that differentiate clusters from each other and build their networking that triggers the performance of the total system further in the human-experience scale.

6.8 Scenarios

The performance of the system as a consistent multi-scalar network is simulated concerning three different scenarios (carnival, weekend evening, and morning rush) while networking triggers bottom-up and top-down reactions at the same time. Space behaves in a different way responding to each scenario. Information flow brings adaptation in urban, island, cluster, and cell scales. The mechanism involves a continuous interaction between networked spatial organizations: a network of cells generating clusters, a network of clusters generating islands, and a network of islands generating the urban proposal. Clusters have the ability to adjust to various scenarios as they are formed by programmatic kinetic cells and responsive canopies. The proper response to each scenario is choreographed according to the input parameters of clusters. Cells through their movements allow the change in the spatial environment that facilitates the development of variable scenarios (Figure 16).

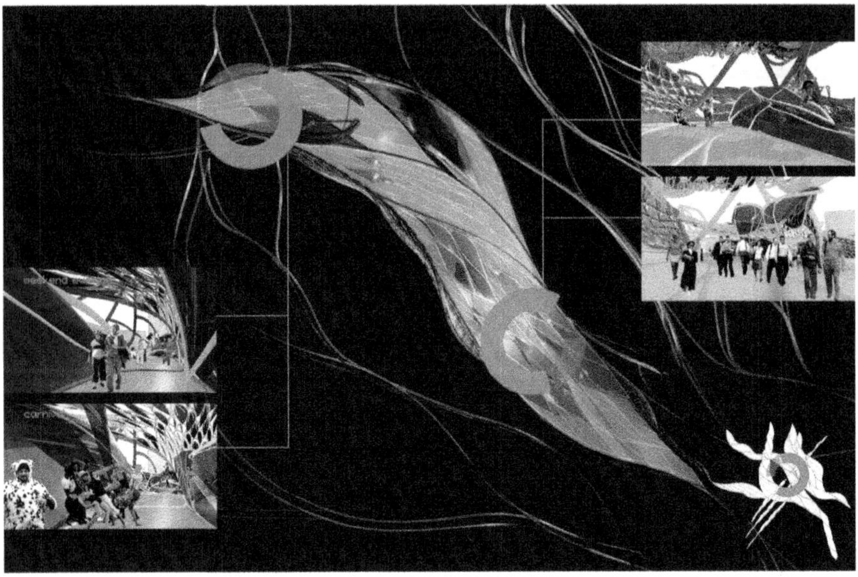

Figure 16. Choreographed scenario in a cluster according to the input parameters of cells.

- Weekend evening at the main clusters of Stratford Bridge: the majority of the programmatic cells are activated creating a landscape for urban leisure. The human interface input is high as users have participated in the creation of their spaces. Crowd activity is intensive and entertainment with community activities prevails while commerce activity is also present on a smaller scale. The self-learning of the system contributes as weekend activities follow repeating patterns.

- Morning rush at the main clusters of Stratford Bridge: most of the programmatic cells are in their rest state; few are activated in order to provide a quick morning passage for the crowd. The human interface input is extremely low. Crowd movement is high. The need for commerce activity is present, while entertainment and community activities are absent. The self-learning of the system partly contributes, recognizing tendencies in the flux of crowds.

- Carnival festival at the main clusters of Stratford Bridge: nearly every programmatic cell is activated creating the infrastructure for a large-scale event. The human interface input is partly activated as some users trigger discrete local events. Crowd activity and movement are high and mostly entertainment with some community activities prevails while commerce activity is absent. The self-learning of the system also contributes as it responds to the need for large, temporary events (Figure 17).

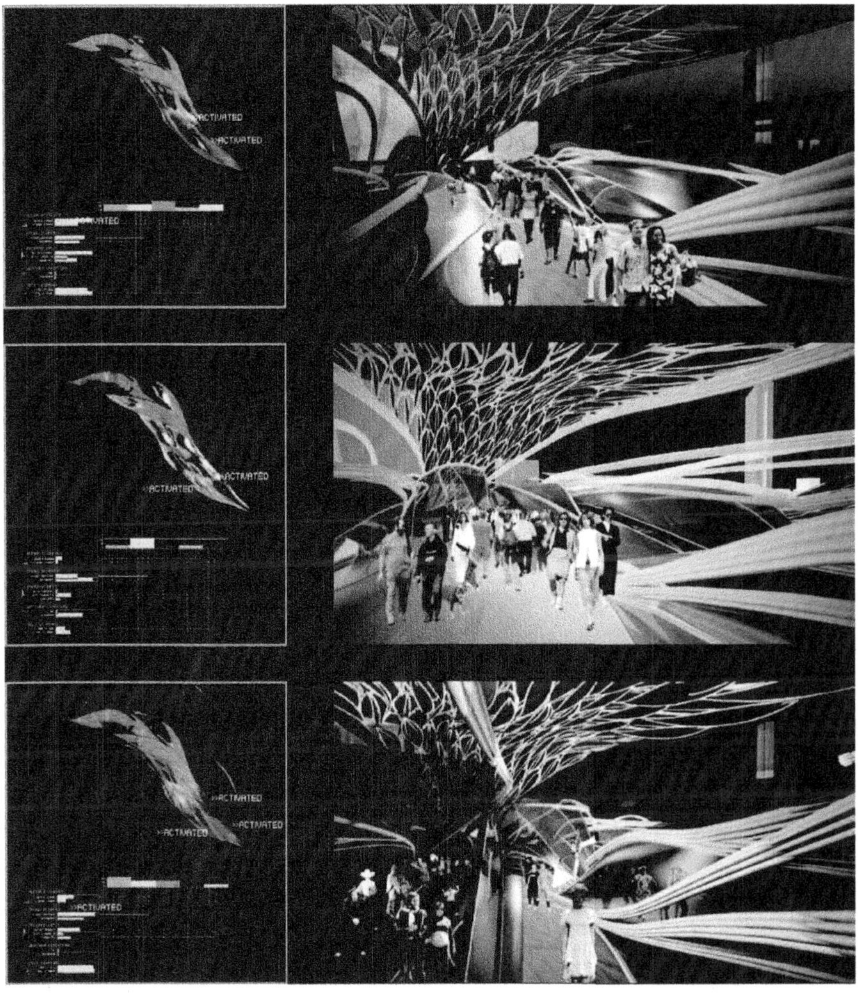

Figure 17. Space (in) formation in different scenarios.

Each island within the urban network is linked together allowing for certain activities to be choreographed among specific groups of islands. Depending on location and input parameters, an island or series of islands can simultaneously provide the same desired activities of a given community.

6.9 Stratford Interface

The understanding of the urban proposal is communicated through the human interface which gives access to users to receive information about the current state of the network across different scales and in parallel input their information (desires of participating and initiating). The networked system becomes open and flexible to accept as further parameters the users' active involvement. The user has an overview of scales, spaces, and time. The urban network behaves like a living organism that shares different information and experiences with users in time and space.

7. Conclusion

The synopsis of the proposal appears at the interface, as each urban island, cluster, and programmatic cell provides the proper information. Navigation through different days of the week and time of the day reveals the flexibility of space and each potential to change. Additionally, information about other users' initiations and preferences could form collective activities and enhance the performance of a networked parametric urbanism (Figures 18 and 19).

At the intersection of the design systems – spatial, informational, and kinetic – responsive spaces converge as a cultural intervention that extends

Figure 18. Prototype of space configuration.

Figure 19. STL prototype of a programmatic cell.

parametric processes to drive the dynamic performance of the urban. This is orchestrated as a system that allows person-to-thing interaction and communication from thing to thing while enhancing the experience of human-to-human interaction.

Acknowledgments The authors thank Tom Verebes, tutor in AA School of Architecture.

Notes

1. Mcluhan.

References

RU3: (2006). Open Network of Collective Intelligence, *RU3: Fuzzy Interfaces, Interaction Design, Information Design*. RU3. 6 April 2009 http://ru3.org/ru3/project/concept/interface.htm

Crandall, J. (2005). Operational Media C-Theory. *Theory, Technical, Culture 32*, (1/6/2005), http://www.ctheory.net/articles.aspx??id==44#bio.

Deleuze, G. and Quattari, F. (1987). *A Thousand Plateaus: Capitalism and Schizophrenia*. University of Minnesota Press, Minneapolis.

Frazer, J. (1995). *An Evolutionary Architecture*. AA Press, London.

Levy, P. (1997). *Collective Intelligence: Man's Emerging World in Cyberspace*. Perseus Books, New York.

Negroponte, N. (1975). *Soft Architecture Machines*. MIT Press, Cambridge, MA.

Chapter 11

THE TOTALITY OF SPACE

THE SPACE OF A BANK AS A COMPLEX SYSTEM[1]

Olga Pantelidou
School of Architecture, National Technical University of Athens (NTUA), Athens, Greece
olga@pantelidou.com

Abstract This chapter addresses a shift in financial spatial thought and a disconnection from contemporary architectural thought and practice. It establishes the concept of the totality of space, an idea and practice unconsciously defined and designed by banking institutions as their business practices and operations evolved due to advancements in communication and information technology, banking law, and managerial theory, through the analysis of the banking industry in the United States. It suggests that architecture needs to address the totality of space and cooperate with such corporations in order to produce conscious, and thus more effective, design results. Ultimately, this chapter addresses the emerging need for a macroscale level of design of a financial institution's space, in parallel to the microscale level, where the architectural tradition is currently focused.

Keywords: architecture, Banking space, Technology, Banking law

1. Introduction

Over the course of the 20th century, a discontinuity of spatial thought between architects and bankers occurred. Architects' efforts were mainly focused on addressing a bank's space at the scale of the unit, while bank corporations began to consider it on a wider scale, corresponding to the totality of their space, which is comprised of a series of spaces that include material and immaterial ones dispersed throughout geography and the connections between them. This expanded space supersedes the primacy of the monad, the building,

W. Minker et al. (eds.), *Advanced Intelligent Environments*,
DOI 10.1007/978-0-387-76485-6_11, © Springer Science+Business Media, LLC 2009

to incorporate the entirety of a bank's space as an open system. Three main factors, banking law, management and operational procedures, and technology, have contributed to how banking thought conceives and, inadvertently or spontaneously, designs its space. These factors frequently intersect. In many cases advances in one prompt developments in the others. Historically, they affect the formation of this space by parameterizing its design on two levels, actualization and control. The result is a new kind of space, a dynamic complex system. Just as its host, the bank, this space is constantly evolving and reforming to address new needs. This space, with its ability to react and in turn affect its environment, can be thought of as an intelligent environment, which is regulated by both internal factors, such as operational procedures, and external ones, such as technology and banking law.

This chapter introduces the concept of the totality of space, defining it as a corporation's bounded spaces and the connections between them. This concept of space, initiated by advancements in science and technology, expresses itself in the evolution of banking in the 20th century. A synoptic view of the history of banking and bank architecture in the United States, along with the factors that shaped them through the 20th century is offered as an attempt to approach a bank's space of operation as a whole. This chapter argues the importance of revealing and understanding the characteristics of the totality of space, which are inherent to the banking industry's spatial thought, thus allowing architects to bring the knowledge of their field and participate in a design/planning process of directing its possible future forms. This implies the need of a macroscale level of design that conceives a bank's space as a whole, in addition to the existing architectural practice of focusing on the needs of an individual building, the microscale level of design. In this manner, architects, specialists in space design, can help corporations purposefully direct the continued evolution of their space as a functioning whole.

2. A Discontinuity

Architecture has always pursued a constant dialogue with technology in order to shape new practices of application. During the 20th century, it focused on new perspectives derived from advancements in construction technology, allowing it to visualize and spawn new forms. These led architects to exploit new design possibilities, preserving and reinforcing the primacy of the building, the shell, the monad. The persistent application of these technical innovations, together with the ambition to rise high, actually drew the attention from new more liberating and fertile directions in spatial thinking.

Beyond construction, technology advanced in other sectors, producing results that were embodied within architectural space products, to support and facilitate everyday activities. The reference here is to communication and information technologies, which from the beginning of their propagation created significant changes in the function and notion of dimension, especially in the social sector of work. Through the constant adoption and implementation of advancements in these new technologies, economic thought together with the notion of work experienced an organic evolution dependent on these technologies' development.

In respect to the sector of work, these technologies directed financial institutions to conceive the space that supports the entirety of their activities as a system. The idea of a system is not new to architecture. A building is a system; a city is a system. In this case, it is a system that is not comprised by a geographically located entity, such as the aforementioned systems, but of geographic dissipation and fragmentation. Nonetheless, it exhibits unifying though complex, chaotic spatial characteristics. Space becomes a totality independent of its dissipation in locus. Not only the constructed boundaries of the shell, but also the physical limits of geography are surpassed. A contemporary vision of space must account for space in its totality. A totality that includes, but it is not limited to multiple buildings, Internet connections, linking through information communication technology, compression of space due to transportation technology, and purely virtual spaces.

The notion of a system itself encompasses the idea of the whole, where the importance of separate units is superseded by the significance of the connections between them. By conceiving of space as a system integral to their operation, banks define a new kind of space, the space between spaces/monads, the space of connections. In fact, the idea of connection is not new to architecture. They have always been a crucial parameter that directs the results of a design. These connections are of a physical nature. While such corporations started with physical connections defining their operations and procedures, they have moved toward a business practice that depends on amorphous, immaterial connections.

The problem that architecture faces in the design and administration of space that accommodates the function of large financial and other multi-national institutions, is located in its preference in handling it as a fragmented whole consisting of separate building shells. Even though architectural thought has already moved toward understanding the space that shelters such institutions as a totality, by starting to address it with corresponding terms, such as "network", architectural practice of design continues to measure it between the limits and boundaries corresponding to the reality of the context of the building. In essence, technology drove banks toward a specific vision of space necessitated by their operations, while architecture focused on construction technology and the spawning of new forms. Architecture needs to conceive of a space design comprised of physical buildings, virtual spaces, and their material and immaterial connections, in order to accommodate and assist the spatial thinking of corporations. It needs to address space as a totality.

3. The Course of Architectural Thought in Banking

The discontinuity between the two perspectives on the nature of a bank's space introduced above is not new. Although, it results from the evolution of banking spatial thought in the 20th century, through the catalytic interference of advancements at multiple levels of technology and management, a similar discontinuity was already present at the beginning of the century in the design of the individual bank building. This impression existed primarily from the point of view of bankers and bank managers. The statement,

> If I had it to do over again I would do more planning myself instead of leaving so much to my architect. I would profit more by the mistakes and success of other bank buildings, instead of relying upon my own judgment. I see now where I could have better fortified my bank's future and at the same time have saved several thousand dollars by pursuing this plan,[2]

made by the president of a well-known bank in the introduction to *Building Equipment and Supplies*, published in 1919 as part of the Shaw Banking Series, reflects the general opinion held by bankers in that era, that an architect could not adequately meet their expectations for their bank buildings. The view that an architect cannot offer a recipe for the success and that the banker himself should research the subject exemplifies the dependency of a bank's future to its space. Although, it addresses the reality and needs of a bank building, elaborating on all the appropriate architectural aspects of the subject from the banker's perspective, *Building Equipment and Supplies* was written with the input of architects. Demonstrating the disconnection between an architect's and a manager's approach to a bank's space, this is one of many books written by members of the banking industry, each of which contains at least one chapter on the design of a bank building.[3]

The period leading to the Great Depression was a period in which the United States experienced a profound growth in banking, accompanied by the construction of many new banks. At this time, America's population was primarily rural, residing mostly in small towns. As such, the majority of banks were in these towns, usually serving as the only bank within a community. In this era, bankers understood that their buildings served as advertisement.[4] Therefore, the expense of a new building could be offset by the revenue it created by drawing in new customers. Even though, large urban centers already supported architects and architectural firms specialized in bank design, at best rural banks only had access to local architects. In most cases, however, they had to rely on builders to design their buildings.[5] The banking manuals described above existed primarily to assist this situation. In some cases, rural bankers ventured beyond their local community for an architect.

The increase in bank construction inevitably led to bank design specialists.[6] They became acquainted with bank practices and needs and offered architectural solutions that effectively reduced the gulf between architects and bankers. A bank could choose between architects who would then supervise construction or firms that both designed and constructed the building.[7] In 1929, Alfred Hopkins, a bank architect, published, *The Fundamentals of a Good Bank Building*, in which, contrary to the voice of many bankers, he states, "bankers do not erect buildings without the advice of those whose task it is to prepare the plans and specifications."[8] Although, this statement is in part an advertisement for his own services, it is true to the reality of the situation, where, there were architects who were in fact experts in the spatial needs of banks at the time. However, not all bankers chose to retain their services. On the whole, the focus of a bank architect was the individual bank building, usually the facilities of a unit bank or a specific branch building for a larger multiple bank. In general, it was rare for bank architects to design large bank office buildings.[9]

During the first part of the century, the bank building sought to establish the image of a monument. Due to numerous bank failures and instances of banking fraud, banks had to convey a sense of strength, persistence, dignity, and security. Their exteriors were reminiscent of Greco-Roman temples and their interiors were decorated with a stern luxury. In general, banks aimed for central locations with plenty of commercial traffic, but were not interested in blending in with their environment. Beginning in the 1920s, banks started to understand that they were a commercial enterprise like any other and that advertising could bring more business.[10] In reaction, the bank building began to change. The bank window became of vital importance, serving both to openly display internal operation and as a space in which bankers could advertise their services.[11] Banks stopped being confrontational and sought to invite the customer. Where the teller's window had previously been a formidable division between the teller and customers, it was made more accessible by reducing the divide.

A central question in construction of a new bank was the type of building to construct, one to simply house the bank's operations or an investment building, which included space suitable as bank headquarters along with space that could be rented to other businesses.[12] Where most rural areas could not economically support the latter solution, urban centers allowed banks to build larger investment buildings. Regardless of what type of building a bank chose, it was vitally important to plan for future expansion.[13] In the case of investment buildings, rented space could be converted to bank operational space as it grew. For other buildings, a bank's expansion could be accommodated by including extra space in the initial design or by planning for additions. Apparently, at least in the first part of the century, most banks did not plan for such an expansion.[14] In cities, the technology of the steel frame allowed banks to answer this question by constructing skyscrapers that housed bank service space, office space, and additional office space which could be rented until it was needed by a bank's expanding functions.

The Great Depression effectively stopped an era of bank building in the United States. It was not until after World War II that banking once again saw significant construction. This generation of banks followed a new business model coupled with a new philosophy of design. Banks became more aggressive in their pursuit of customers. Through their design, they sought to attract customers directing them to new services. To this end, modernist architecture, with its emphasis on clean lines, offered glass facades, a new possibility in construction. New banks boasted inviting entrances that lifted the secrecy, previously held as important. Inside, a bank's public space was open and inviting, directing customers to various services.[15] As Charles Belfoure states in *Monuments to Money*, "many banks across America used the same combination of the glass box within a concrete frame for their branches. . ."[16] Following the desire to create more inviting spaces, banks also updated their interiors, taking advantage of the introduction of fluorescent lights and advancements in air conditioning. Both technologies aimed toward a more pleasant environment. Where fluorescent lights countered the foreboding dimness of the previous era by creating a uniformly brighter interior, air conditioning regulated a branch's temperature, keeping it comfortable year round. This period also experienced a new form of bank expansion. Where the previous era focused on expansion in urban centers and rural areas, following World War II, banks followed the American population into the suburbs, usually placing new branches in commercial and shopping centers.

The last quarter of the 20th century experienced a reversal in trends concerning the appearance of banks. Bankers returned to images of dignity and strength. For smaller buildings, this resulted in forms reminiscent of the beginning of the century. In large cities, banks expanded the skyscraper's role from investment building to icon.[17] This period emphasized on the notion of a bank as a retail business, whose ultimate goal was to sell products. This is evident in the appearance of supermarket branches and in banks' activities related to branding. Although, supermarkets had been a part of the American landscape from early in the century, it is only after they offer retail spaces within their premises that banks could take advantage of them. In contrast to traditional bank branches, a supermarket branch offers lower operating costs and immediate access to potential customers.[18] The concept of a bank space as a retail one had a profound effect on the interior design of branches. Banks that embraced this concept created interiors that borrowed techniques from the retail industry. The widely varied results, while different from traditional branch interiors, continue to direct customers toward products, while engendering a feeling of great service.[19] This period also witnessed the emergence of a unified bank brand. Although, the concept of branding already existed within business practices, it was not until banks went through the series of mergers and acquisitions, permitted by changes in banking law in 1994, that branding became a necessity in bank design. Faced with branching networks that spread through out the United States built out of multiple banks, bank holding companies understood the importance of creating a unified customer experience, both to function as a whole and as a way to increase customers' awareness of their products. In order to accomplish this goal, banks started to require that each branch follow specific rules in designing their interiors and exteriors.

4. The Course of Banking Spatial Thought

Concurrent to the evolution of a bank building's architecture, bank management and organization was going through its own development. While most bankers were content to operate out of single facilities, and in fact resisted other trends, some felt urged to explore new directions of bank operation.[20] They found that bank operations based in a single building limited their business' potential. Instead, they sought to expand their space from one building to include multiple spaces in various geographic locations. Banking law was the pivotal limiting factor in determining not only the manner in which banks could function as business entities, but also the manner in which such institutions could expand geographically. The shape of a bank's operational space at the turn of the 20th century was primarily regulated by the National Bank Act of 1863. Prior to 1863, a period known as the Free Banking Era, a number of states allowed the automatic chartering of banks assuming certain minimum requirements were met.[21] During this time, there were no national banks and each bank had the right to distribute its own currency. The National Bank Act of 1863, which established a national banking system and a unified monetary system, is the first comprehensive effort in the United States to define banking at a national level. Interpretation of this act limited national banks to "one brick and mortar office"[22]. Even though the act itself did not actually contain this language, later amendments strengthened this interpretation.[23] As such, national banks were limited to unit banking. The Unit Banking System was the traditional form of banking in the United States. This system is defined as:

> ... a banking system in which each banking office has its own separate corporate organization and is owned and managed as a separate institution. A unit bank has only voluntary correspondent relationships with other commercial banks.[24]

Most banks, including all national banks and nearly all state banks, operating during the first quarter of the 20th century were organized under this system. Basically, financial institutions operated as independent legal entities that occupied one building, which accommodated both customer service and their other functions. The unit bank operated as an entirety within the monad of its building. To expand their business to other locations, they created correspondence relationships with other banks. It should be noted that certain banks continue to organize themselves under the unit banking system in the United States.

The Bank Act of 1863 left the nature of intra- and inter-state banking to be determined by the laws of states themselves.[25] For banks chartered at the state level, the different types of laws enacted by states resulted in two types of banking, unit and multiple. The latter form of banking refers to financial institutions in which numerous banks were owned and operated by the same individual or holding company. A more complete definition is offered by Virgil Willit in *Chain, Group and Branch Banking*,

> Multiple Banking is a comprehensive term which includes any system of banking in which two or more banking offices, performing the fundamental commercial banking functions of receiving deposits and making loans, are subject to common ownership in whole or in part, and to common control. This ownership and control may be vested in one person or a group of persons, a bank or trust company, a holding company or other corporation. The term thus includes branch, chain, and group banking.[26]

Where unit banks were limited to just one building, a bank that operated under a multiple banking system existed as a unified legal entity across multiple buildings in separate locations. The exact arrangement of its locations was limited by state laws. In states that allowed limited branch banks, branches were required to operate within specific geographic limits of their main office, while in those that allowed the practice of multiple banking, a bank could establish branches throughout the state.[27] In some cases, there were also limitations on the type of additional space that a bank could operate, for instance, whether it was a branch with full services or an additional office with limited functions.[28] Therefore, even at the beginning of the century the overall shape of a bank's space as a whole was directly dependent on the banking system to which it belonged and the state laws under which it operated.

The term multiple banking primarily described three distinct banking systems: branch, chain, and group banking. From the bankers' point of view, all of the three systems were means for physically expanding their operations geographically, while confronting restrictions posed by banking law that dictated geographic limitations. In branch banking,

> ... two or more banking offices are maintained by the same banking institution. These offices or branches of the parent institution perform all the usual banking functions but do not have a separate corporate organization. They are managed by a manager appointed by the parent institution.[29]

Comptroller of the Currency from 1928 to 1932, J.W. Pole, defined chain and group banking as follows:

> The term "chain" ... describe[s] a condition in which a number of banks were owned or controlled by the same individual or by a group of individuals. These so-called chains were situated very largely in the rural districts and the members of the chain were principally small country banks ...
>
> The term "group banking" ... appears to be a major movement in our banking system. The principal factor in group banking is that each group is centered around a city or metropolitan bank through means of a holding company, which owns the majority of the stock of each bank, thereby creating a system of banks more or less integrated in management with the central bank of the system ... The principal difference between a group and a chain is that the group always has some form of central management while the chain has not.[30]

The "major movement" in banking described above refers to the trend of group banking started by Bank of Italy, the forerunner of Bank of America, which sought to find a way around laws that limited chain banking in California.[31] The result was a new, legal, unlegislated form of multiple banking, group banking.

The enactment of the Federal Reserve Act of 1913 began a period during which the exact nature of national banks and their branches was formally established. Among other things, it required that all national banks become members of the Federal Reserve System.[32] Under this Act, a national bank could retain any branches owned by other banks it absorbs. It also provided an option for qualified state banks to join. If a state bank converted to a national bank, it was allowed to retain and operate its branches.[33] With this Act, national banks were finally, formally allowed to operate branches under certain conditions. They still, however, did not have the same privileges enjoyed by state banks in establishing branches.[34] The McFadden Act of 1927 further placed limits on branching for both national banks and state banks that were members of the Federal Reserve System. Such banks were limited to branches within their home city. Thus, the McFadden Act effectively prevented national banks from crossing state lines.[35] During this period, banking law underwent corrections, regarding the branch banking question, in an effort to equalize space privileges between different types of banks, namely national and state banks.

Beyond the two aforementioned types of banking systems, unit and multiple, two types of bank holding companies emerged: one-bank holding companies and multibank holding companies. The former enabled branch banks to possess companies, whose activities were closely related to banking, thus allowing them to offer more services. The latter permitted unit banks to circumvent restrictions on branch banking practice.[36] The strategy of the holding company was invented to enable banks to gain greater space advantages than allowed by banking law at the time. Initially, there was no legislation that specifically addressed bank holding companies.[37] Instead, such corporations were often challenged by anti-trust laws, having repercussions on their space, making them a risky venture.

The next important shift in banking law, during which the exact nature of holding companies was addressed, began in 1956. Prior to this, it was unclear whether the Federal Reserve or the Justice Department was responsible for monopolistic practices within banking. Because a clear definition was not established, few banks were willing to form holding companies. The Bank Holding Company Act of 1956 clearly delineated the nature of such corporations. Beyond placing restrictions on the activities in which a bank holding company could engage, it prevented such corporations operating in one state from acquiring a bank in a second state, unless authorized through statute by the second state. This Act also exempted bank holding companies that existed prior to its enactment, by allowing them to have non-bank affiliates in other states.[38]

The nature of bank holding companies was further clarified with the Bank Merger Acts of 1960 and 1966. These acts clearly defined how two banking institutions could merge lawfully. These three Acts alleviated the fear previously associated with forming a holding company, allowing more banks to adopt this model as a viable strategy of growth.[39] This third phase of banking law permitted an adequate degree of flexibility, practiced through nearly the end of the 20th century by banking corporations, in the distribution of their space as a whole. The Riegle-Neal Interstate Banking and Branching Efficiency Act of 1994 further lifted restrictions on bank branching. According to it, from 1997 onward, banks could purchase banks in other states, regardless of state laws restricting such activities, assuming that the resultant bank did not control more that 10% of deposits in the United States and no more than 30% of deposits in the state, unless state law specified otherwise.[40] This Act effectively allowed bank holding companies to create bank networks far larger than previously possible.

The 20th century witnessed banking evolving from a predominance of unit banking to a prevalence of holding companies operating under the multiple banking system. The space of these institutions can no longer be comprehensively described through the list of buildings they occupy. Instead, it can only be described as the network of its subsidiaries and the space they inhabit. While bankers focused on the operation and design of their emerging banking networks, architects were enlisted to design portions of the system, usually one building at a time. Even in cases in which architects designed multiple branches that followed a bank's brand, they were only addressing a small portion of something far larger and more complicated. The result is that financial institutions function in an "unplanned" complex space that is not delineated by specific buildings. The form of this space is not consciously designed. At times market analysis lead to choices regarding its expansion, but more often, mergers and acquisitions lead to the union of multiple spaces that had originally evolved following different sets of strategy and tactics. In the end, the nature of a bank's overall space, to a great degree, emerged due to its management and history.

5. Technology's Effect on Banking Spatial Thought

During the span of the 20th century, financial institutions developed a close relationship with technology in order to promote advancements in their operation and administration. Traditionally, they exhibit conservative behavior in adopting change in contrast to other types of corporations. This conservative behavior regarding change exists regarding technology as well.[41] Nonetheless, from the beginning of the 20th century, they recognized that technology played a crucial role in the execution of their work in terms of efficiency, accuracy, and effectiveness. During this time, a series of technical developments appeared, promising more advantageous bank operations, affecting a wide variety of functions. A letter, for instance, could be dictated to a recording machine, which was then given to a typist for playback and transcribing, reducing the need for a stenographer. It could then be copied either using a copier or a letter machine. The latter method was often preferred, even though it required typists to personalize salutations and addresses, creating the illusion of personal communication. Next, multiple copies were sent to addressing machines, folding machines, and envelope and stamping machines for mass mailing. Communication was also extended through the use of the pneumatic tube and telephone. At the beginning of the 20th century, the pneumatic tube was important for the interoffice communications it allowed. The phone, although seen as a modern tool that could allow better interdepartmental communication, was not widely used, for fear that it would create misunderstandings.[42]

This era also saw advancements in information technology. Mechanical calculators, known then as computing machines, allowed an increase in accuracy and speed of computations, while counting machines permitted larger amounts of data to be processed. Procedures were implemented around uniform forms for most activities within a bank. Based on the operation, forms often had specific colors. To further expedite the process, multilayered forms were created that allowed only certain parts of the information to be copied to lower layers. Each layer was a different color making it easy to identify which one went where.[43] This is the beginning of standardization of information and bookkeeping across multiple departments. This was accompanied with new approaches to filing information, which evolved from single hard-bound books to binders of loose paper, to flat files, and finally to vertical files with tabs that made it easier to identify sections at a glance.[44]

From the beginning of the 20th century, the choice of the proper technical equipment had significant repercussions on the image of a bank, since the notion of "modern" was already deeply entrenched in the practice of banking. As early as 1934, the American Bankers Association notes in a booklet on commercial bank management that, "the major fundamentals which a bank must possess before it can hope or deserve to grow are *strength, clean history, personality, adequate equipment, and satisfactory service*."[45] The importance of technology in the routine work of a bank is characteristically outlined in the following statement:

> The necessity in the modern bank for speed and precision and legibility has brought into being scores of machines which assist and supplement hand labor, ... Then also there are all kinds of ingenious and practical forms which minimize time and facilitate service. ... Many of the larger banks have pneumatic tube service between the various departments so as to make possible quick and safe interdepartment communication.[46]

During this time, the cutting edge equipment that a bank could bring to its operation is stated to be "... bookkeeping machines, posting machines, coin counters, automatic change making machines, envelope sealers, envelope openers, typewriters, adding machines, machines for writing checks and machines for canceling them."[47]

These technologies ultimately allowed banks to expand their customer base, both in number and in distance, allowing them to extend the scope of their operations. Some technologies enabled the offering of new products, such as banking by mail, which was dependent on mailing, computational, and filing technologies. Technology supported correspondence between different banks by equipping them to become more organized, clearer, and faster. A bank's business hinged on information, especially credit information, collected from networks that included mercantile agencies, local correspondence, traveling salesmen, merchants' associations, and credit reporting agencies.[48] As volume grew, technology increased banks' ability to collect, process, and file information. Such innovations were key in creating the circumstances necessary for better customer service to more clients. Technology's effect on space is seen through the growth it promoted. Larger customer bases necessitated better communication and better information handling. The selfsame technology enabled banks to grow to a point that required new forms of organization. In the case of more traditional banks, this lead to departmentalization of operations. Whereas more aggressive banks, already present at the beginning of the century, moved even further toward the invention of branch banking practices.

Bank of Italy, for instance, took advantage of this situation to expand through branching, mostly in rural areas, which up to that point could not afford the luxury of bank installations, usually due to their remoteness from large urban centers.[49] Technology offered fertile ground for conceiving distance through a sense of proximity rather than "distance." As such, it offered an unprecedented opportunity to create an early stage of coherency between spaces dispersed to a limited, though considerable for the time, area of one state.

The period that follows is greatly influenced by the general adoption of the automobile into daily life. The car had the effect of reducing distances, expanding the geography from which customers could reach a bank, and, equally important, easing communication and connection between a bank's branches. Moreover, many financial institutions were motivated to accommodate this technology by creating appropriate new spaces, drive-through tellers. Furthermore, customers' newfound mobility created demands for a unified branch experience in a bank's space.

In addition, the most forward thinking banks established departments of research to explore ways in which technology could both improve products for their customers and enhance their own internal operations. This period witnessed the influence of a new technology, the computer, known at that time as the "electronic machine," which continued the trend of expanding volume and automation. Quite indicative is the case of Bank of America. As one of the most progressive banks at the time, it integrated its operation with new information and communication technologies that radicalized the whole process of its workflows, as well as its service to customers. Bank of America had already created departments dedicated to the research of technological innovations, the Systems and Equipment Department and the Customers Services Department. The first was charged with the development and application of technological methods to reduce costs and improve the efficiency of the bank's operation.[50] Its research was focused on five areas: premises, methods, computers, communications, and equipment. The second was assigned the task of translating technological change into new services for the convenience of their customers and into additional sources of income for the bank. As noted in the 1960 Bank of America's Annual Report, this department studied ways in which the facilities of the bank could be adapted to better serve business, while new services for customers, such as Freight Payment, Utility Billing, and Customer Payroll services, were constantly being introduced.[51]

At this point in history, even though there was a greater trust of the telephone, communication technology had not evolved greatly. Accounting technology, on the other hand, expanded by leaps and bounds, allowing a profound increase in the number of customers. Bank of America is credited as the first bank to introduce electronic accounting, now known as computing, by implementing ERMA (Electronic Recording Method Accounting) in 1958.[52] The relationship that evolved between the bank and this system is indicative of what has become standard operational procedure between information technology and financial institutions. Initially, the system operated in two centers handling the electronic bookkeeping of 200,000 customer accounts. Within a year it expanded to nine centers. One year after that, it grew to 13 centers serving 90% of Bank of America's branches. At this point, the system extended its accounting capabilities to include customer accounts, traveler's cheques, and the internal sorting and handling of Bank of America cheques. Through the use of daily air messenger service, all but 22 geographically remote branches were serviced by Bank of America's ERMA and Data Processing Centers. It is important to note that information traveled physically by land or air to accumulate in the bank's Data Processing centers. Ultimately, through this system, Bank of America was capable of processing 2.5 million checking accounts.

During the early 1960s it had computer systems for handling loans, checking accounts and traveler's cheques. Concurrently, it also maintained additional computer systems that handled the accounting of other parts of its operations, such as cheques to and from other banks. The scale of banking operations allowed by this kind of technology also necessitated its use. This exemplifies how standard bank practices became dependent on new information processing technology. Demonstrating the speed at which this technology transformed and the need to maintain the cutting edge, in 1966, Bank of America abandoned its 12 now obsolete ERMA centers and replaced them with two IBM 360 Model 65 systems.[53] The statement from their annual report,

> Although this conversion effort has been fraught with the problems common to the application of advanced technology, we are confident that, as one of the first commercial users of this new third generation equipment, we are not only preparing for substantial future cost reduction but also are preserving our technological lead in the application of electronics to banking,[54]

indicates how quickly even at the beginning banks understood the need of maintaining their information technology at cutting edge levels. Their dependency on this technology, along with the economic gains and the opportunities of growth it created, outweighed the cost and the effort of constantly remaining up-to-date.

The adoption of electronic accounting systems introduced a nearly universal unification of a bank's pool of information. Since information is the fundamental basis of a bank's function, this had a profound unifying effect on its operations. This technology not only grew out of a developing need for everexpanding bank activities, but it also created the necessary conditions that allowed this expansion. These systems affected a bank's space in two ways.

Most directly, they enabled a bank to spread further geographically, reaching even more remote locations. Even more importantly, they created more direct connections between each of a bank's spatial entities, by centralizing all of its information, and then distributing it as necessary to its dispersed loci. The advent of such technologies initiated the process of removing blockages imposed by distance, while concurrently allowing more seamless interactions among the various physical spaces of a financial institution. When all the parts of the whole enjoy the same advantage of communication and connection, the importance shifts from the manner in which parts are connected to the mode in which the whole operates. As such these technologies reduced the effects of the physical isolation of a bank's branches, creating at certain levels stronger connections independent of distance.

During this time, financial institutions also introduced new products allowed by other advancements in technology. These products, such as credit cards, had catalytic effects on their clients relationship with the bank's space. Bank of America serves as an example again with the launch of its first all-purpose credit card the BANKAMERICARD, which coincided with the introduction of electronic accounting.[55] Credit card technology profoundly impacted a customer's relationship to a bank's space. It granted customers more independence from a bank's physical space, allowing them to engage in banking operations away from bank premises.

In part, the final third of the 20th century saw the continuation of technological trajectories established earlier in the century. However, this period also experienced advancements in technology that led to new methods and solutions concerning banking practices. Following the previously established paths, advancements in technologies like the computer continued to increase speed and efficiency of data processing, now encompassing the sum total of a bank's information. The introduction of the personal computer replaced countless mechanical predecessors and modified various bank procedures and filling systems. The ease with which large quantities of information could be obtained and processed greatly increased a financial institution's ability to analyze and manage risk. In these examples, the computer serves to update technological trajectories already present in bank practices.

In addition, banking underwent two other significant changes in technology, through new forms of automation and connectivity, which created new trends in the nature of their expanded space. The example of the automated teller machine (ATM) embodies both trends. In their initial introduction, ATMs were only available to trustworthy customers with good credit. This limitation existed, because money drawn from an ATM functioned essentially as a loan. The original ATMs were incapable of accessing customer account information. Money would be subtracted from a customer's account at a later time. Even though their use was limited, the original ATMs introduced a new space of service that was accessible to customers 24 hours a day, expanding a bank's space both physically and temporally.

ATMs also illustrate the repercussions of network technology on a bank's space. The culmination of technology in the 20th century was the union of information and communication technology. Through it, not only could a bank store and process countless transactions, but also it could make this information available at any point within its network nearly instantaneously. The most important effect of networks on banking is the immediate interconnectivity between every point within a bank's system. In the case of ATMs, network technology made it possible for financial institutions to extend this service to all of their customers. A cash withdrawal was no longer a short-term loan, but an actual debit from an account, and money could only be accessed if enough was available within an account.

Furthermore, network technology offered novel possibilities for additional services through ATMs, expanding this new space's ability to serve customers even further and allowing customers more freedom in their relationship to a bank's service space. Of course, banks as conservative institutions were initially skeptical of the adoption of ATMs. In 1975, Citibank gambled on the yet unproved concept investing an unprecedented $100 million and by 2 years into the development of its first ATMs. In 1977, virtually over night, it placed 400 ATMs in New York City.[56] This demonstrates the manner in which technology enabled a bank to dramatically expand its space of operation nearly instantaneously by conceiving and implementing a completely novel service space. Furthermore by 1981, the introduction of ATMs doubled Citibank's share of demand deposits in a previously stagnant market,[57] creating a potential for further expansion of Citibank's space through additional branches and even more ATMs.

The distribution of images through a network, first considered as a tool for increasing efficiency in banking operations by Chase Manhattan Bank in 1993,[58] is another example of a bank technology that transformed banking procedures and extended a bank's space. In this technology, images refer to digital reproductions of bank documents. An electronic image of a document could be made available at any node in a bank's network instantaneously, whereas a hardcopy document had to be physically transferred from one location to another with much more limited speed. Chase was the first to understand that this technology translated into continuous processing of documents. A branch in New York could start processing a document, and then, at the end of its day, when it closed, an office in Hong Kong, which would just be opening, could continue processing the document.[59] With this imaging technology, over the life of a document, a bank's space is continuous across geography and operates 24 hours a day. This phenomenon, first experienced by customers through the adoption of ATMs, was thus extended to a financial institution's internal operations as well.

Internet banking is the culmination of trends started by the introduction of ATMs and other network technologies. ATMs created customer expectations of even greater independence from the bank itself and of 24-hour service. Network technologies suggested that bank transactions did not necessarily require, and in fact might be limited by, physical documents. The introduction of secure web browsers to the already popular World Wide Web opened the possibility for Internet banking.[60] Although banks had already adopted internal networks, secure web browsers allowed them to share information with customers anywhere an Internet connection was available, ultimately resulting in online banking. Like the ATM, online banking enabled customers to conduct bank transactions at any time, but in an immaterial space, beyond a bank's material service space. Unlike ATMs, online banking can occur anywhere, extending a bank's "transactive" space into a virtual space represented by its web page.

A successful bank's relationship to technology can be summed up by David Rockefeller's 1964 statement, "I think of creative banking as a willingness to re-examine all segments of our business in the light of the changing technology of banking, but as an unwillingness to be bettered by arbitrary rules and routines."[61] The beginning of the 20th century witnessed financial institutions that were already dependent on available technology for the successful execution of their procedures. Technology provided banks with increases in efficiency, accuracy, and speed, while simultaneously decreasing the effects of distance. As a result, their customer base and activities expanded, necessitating further adoption of newer technology in order to accommodate additional growth. These technologies had an immediate impact on space, as it was designed to house the technology itself. Technology's ultimate effect, however, is the collapse of space into a continuity. Where older methods of communication and transportation exposed physical boundaries and distance, new information and communication technologies introduced immediate feedback regardless of distance. The result on spatial thought is to allow a financial institution to operate in a continuity of space independent of its physical spaces' geographic distribution. Of course, changes in technology and space also create new customer habits and expectations, which then in turn demand further evolution of technology and space.

6. The Contemporary Reality of a Bank's Space

In order to fully understand the new direction in architecture proposed here, an in-depth understanding of financial institutions' operations is necessary especially as those operations apply to the use of space as a whole. The information presented here only offers the glimpse necessary to introduce this idea. This allows an understanding of how large institutions conceive of space and how such space can be designed. As such, large financial institutions, such as Bank of America, Wells Fargo and Chase Manhattan Bank (known as JP Mogram Chase from 2000 onward) whose operations have spanned the entire 20th century, offer a fertile ground for exploration. During their history, these corporations constantly seek to maintain a modern image by embracing and implementing cutting edge technologies and through the pursuit of innovation, either by being innovative themselves or by buying companies that exhibited special innovative characteristics in their operation, a fact that has repercussions on their conception of space that supports their function. Such corporations essentially consist of an assemblage of numerous, varying companies, whose products are immaterial and therefore their activities are limited to office and service spaces, which serve as their space of production and distribution respectively. Although, they began as companies with specific content, their history of acquisition has transformed them into holding companies that embrace the whole field of finance. By employing mechanisms of selection, extinction and generation, they experience events of growth, contraction, and reshuffling. These same events and mechanisms engineer their concept and use of space, which transforms with each critical choice. In this manner, such corporations inadvertently and spontaneously design their architectural space, by synthesizing shells and spaces designed by architects, into a coherent whole.

Throughout the 20th century, the space of these corporations changed constantly. Shifts in their operational needs, in fact, can only lead to transformations. Such shifts often occur due to variable factors of the standard business practices. Among them one can count events such as expanding client base, expanding product offerings, and reactions to changes in the marketplace. The resulting product of such business acts corresponds to space remodeling, either by the addition or subtraction of space entities through the construction of new buildings or the purchase and selling of existing ones. The constant adoption of new technology often required new space models of function. Initially, technology only required changes in space design and configuration, in order to be incorporated into the working process of an office building's context. As technology evolved, novel modes of business practices emerged, having even more profound effects in space usage and exploitation, and moreover, in its notion for work accommodation. Acquaintance with such technology leads to experiencing dispersal of workflows. The convenience of dissipation in the locus of work production challenged the embedded notion to encompass even homes as ordinary workstations. Current trends in information and telecommunication technologies have allowed the space of work to be a malleable, flexible, limitless environment that manages to maintain the key characteristics of unification and continuity. Space also experiences changes due to acquisition of other companies or the disposal of subsidiaries, a fact that leads to the growth or the contraction of space itself. Through such processes, companies experience changes in their identity profile by exchanging qualities and characteristics with the other(s) process participant(s). As a result, their space of action experiences corresponding alterations, dispensing with familiar attributes or inheriting evolving properties.

With each alteration, changes in operations occur, often involving the movement, division or reallocation of existing headquarters, branches, key offices, and departments of the corporation. These are just some of the instances that translate into changes in a company's power character, expressed by shifting from hierarchical to hegemonic models of administration and governance. Whereas the well-known hierarchical models follow a top-down approach to decision-making, the already emerging and implemented hegemonic models foster decisions at multiple levels, but with centralized oversight. Hegemonic modes of operation embody the level of flexibility in administration accompanied by "new forms of centrality," an idea proposed by Saskia Sassen,[62] necessary for operations to expand to massive scales and globalization. As Sassen points out in her essay "The global city: strategic site/new frontier," "the geography of globalization contains both a dynamic of dispersal and of centralization, a condition that is only now beginning to receive recognition."[63] Here, two key general characteristics of corporate space are recognized. The first is dissipation in locus embedded with new forms of dynamics in action and interaction. The second is identified as a new kind of centrality that is now recognizable as cores of administration dispersed to cities that serve as global centers.

Nonetheless, these institutions present a picture of solidity in the reliability of their operations. In fact, they are never at equilibrium. They are constantly evolving in order to remain competitive, successful, and up-to-date. Although, they are not rigid, they manage to achieve states of stability, progressing to higher orders of complexity. As such, they exhibit characteristics of self-organizing entities found in complexity theory. The space of their function is a resource for the company that combined with its operation forms a relationship of organic growth (or contraction) and/or adaptation/evolution. Even though space is still literally described as space has always been described, when discussing their organization, operations, and procedures, this relationship suggests that corporations unconsciously define the description of their function space as a totality dependent, of course, on the material, but especially on the immaterial connections of their function and organization.

Historically, these are institutions that moved and operated through various organizational/managerial models; allowing them to remain up-to-date. Each adopted philosophy has brought about shifts in understanding of space, moving beyond the local, microscale level of design, the building, toward a (though not yet clearly acknowledged) macroscale level of design, the totality of space. This shift from a need of microscale design to macroscale design occurs as such companies conceive of themselves as networks, and not disparate units of operation, dependent on connections, which become more and more immaterial with the advance of technology. This philosophy of thought in operation and space thinking is the result of the economic theories that gradually shaped the reality of the 20th century. In general, these economic theories were deeply influenced by, or even conceived due to, advancements in scientific fields such as mathematics or physics and the nascence of new theories such as chaos theory. As a result, the shape of these companies, which is, in fact, translated into their operations and the space that shelters them, is now modeled through the language of mathematics. Therefore, space should be considered as an entirety. It can no longer be described through Euclidean maps but by the multi-dimensional and multi-variable mathematics of physical systems found in complexity theory. At the end of the 20th century, a bank is seen as a complex system, whose space is most directly affected by three factors banking: law, bank management and operational procedures, and technology. To understand the space of a bank as a whole, the relationship of space to a complex system, as well as the effects of these three factors must be analyzed.

7. Space of a Complex System: The Totality of Space

Even from the beginning, financial institutions that were organized as unit banks functioned as systems, especially since their activities expanded to require organization into departments. The shift to multiple banking expanded the notion of such institutions as systems; the system embodied by one bank building expanded to include additional new units in the form of branches, additional offices, ATMs, etc. As it grew, the organizational complexity of a bank reached a point that it could be considered as a complex, open system.

Space is the container body of the system that it serves and accommodates, comprising an integral part of it. For a complex system, such as a bank, space by necessity refers to its space as a whole, which includes all material and immaterial spaces. Among other things, a system's space can be used to visualize its boundaries. As this is an open system, one which is dependent on a constant flow of information, energy, and/or material with entities beyond its boundaries,[64] identifying these boundaries is necessary for describing the system. The totality of space also helps to comprehend the internal structure within a system. If space is defined purely through physical characteristics, such as the boundaries of buildings, the system exhibits internal boundaries that require bridges. These bridges are constantly invented through the implementation of specific media, i.e., operational procedures and the technologies that support them. By overcoming these internal boundaries, the totality of a system's space emerges. The totality of space is defined as a system's bounded spaces and the connections between them, both of which may be material or immaterial in nature. It is limited to the space that is exclusively and directly controlled by an institution. Ultimately, the totality of space encompasses the spaces and connections on which a corporation depends on to support its operations and procedures.

A complex system cannot be understood through the reductive practice of summing the understanding of its parts.[65] The non-linear nature of such a system is exhibited by the fact that changes in any of its parts cause non-proportional changes throughout the system. The whole can only be grasped through understanding of the overlapping connections and relations throughout. For this reason, the totality of space only allows a glimpse at the form of the system it contains. Instead of describing, it reveals certain aspects of the system's formation. For instance, awareness of every type of space and connection informs regarding the system's organization and operation. Moreover, operations and procedures play a crucial role in forming a system's space. This is just as true at the level of a spatial unit, as it is for space as a whole. For this reason, upon entering a branch, space makes certain of a bank's procedures, such as its dual goals of friendly customer service and security, obvious, on the one hand directing toward desired services, while on the other delineating protected areas.

When viewed from the totality of space this effect is repeated revealing more, though abstract, information concerning the system as a whole. Its spatial nodes are identified along with their types, distribution in multiple loci, and modes of connection. A description of the totality of a bank's space reveals much of a bank's current organization. The example of Bank of America's arrangement in space illustrates this point. This bank currently has its headquarters in Charlotte, North Carolina, its Information Technology headquarters in Los Angeles, its office overseeing its international branches divided between Los Angeles and San Francisco, branches spread out throughout the United States and internationally, and operational offices, which incorporate a variety of operations in cities throughout the world. This brief and incomplete description of a portion of Bank of America's totality of space already suggests aspects of the bank's operational organization, revealing its center of operation, its peripheral centers, the relationship between the various parts and the whole, and their dependency or independency to other parts of the system.

The totality of space can be thought of as one layer within the network that describes the entire system of a bank. It is a subsystem of the institution that when treated as another complex open system, whose relationship to the bank is described through exchanges at its boundaries, gives a few additional insights. The definition of the totality of space, taken to include a bank's bounded spaces and the connections, does not necessitate complexity. Complexity in this expanded space occurs through iterative growth, initiated and exercised by its host, eventually resulting in even higher levels of complexity.

This iterative process assumes constantly changing parameters, including environmental constraints (i.e., population density, local laws, etc.) as well as new forms of space (i.e., office buildings, IT centers, drive-through tellers, ATMs, online banking, etc.) and connections (i.e., information communication technology, such as mail, telephones, LANs, and the Internet, and transportation technologies, such as automobiles and airplanes) that due to non-linearity in results lead to unpredictable forms of the totality of space.

The case of Bank of America again offers a fertile example, quite typical of this process. It embodies the growth of a single space to a complex network of spaces, by following a single process of addition of nearly identical entities. Originally, in 1904, this corporation started as a single unit bank in California, under the name, Bank of Italy. Quickly it adopted a strategy of growth through the acquisition of numerous other unit banks within California. The act of acquisition serves as the iterative process. Between 1904 and 1926 it grew from one branch to 98.[66] Here, variation in acquisition is essentially limited to the location of the new branch and its pre-existing customer base. Nevertheless, the number of acquisitions resulted in a totality of space consisting of nearly 100 branches and a complex, though technologically primitive, network of communication that could not be predicted a priori.

During this period, under the same leadership, in response to legal restrictions, three other separate legal banking entities were also created. In 1927, changes in banking law allowed these four entities to legally merge into one under the name Bank of Italy.[67] Two years later, in 1929, it experienced a new form of growth through merger with Bank of America, organized under the latter name.[68] Two totalities of space were brought together, forming a new one, whose characteristics were not simply the sum of the previous two spaces. Although, the sum total of the bank's new physical space can be described as such, the resultant totality is at very least formed through the combined physical space, the two sets of previous connections and the new connections necessitated by the merger. Ever since, through a continuing practice of mergers and acquisitions along with continuing changes in types of spaces and connections, this bank's space continues to grow in unforeseeable manners.

The totality of space emerges from a series of repeating procedures in which the variables of each repetition are not constant. Their nature is such that even small variations can lead to great differences in the outcome. For a bank, these procedures can refer to an opening of a new branch, or a merger between banks. In the example of a new branch, variables in branch opening procedures, such as the cost of buying, renting, or building the branch; the size of branch necessary to serve its location; the services offered by this branch; and the means of communication between the new branch and the whole, affect the specific form of the new space and, by extension, the change to the totality of space. Although, this change in and of itself may be predictable through limited repetitions, when repeated numerous times the results lead to unpredictable forms of space as a whole, which in turn affect the system in unpredictable ways. This is the process that leads a system to greater levels of complexity.

Although space itself is inert, the totality of space, which includes both inert space and the connections between, can be thought of as a "living system," as its connections create a state of ongoing flux and change, allowing it to be considered an open one. Its environment, with which it interacts and is dependent on, exceeds that of the bank's system to include the latter's external environment. As such, the totality of space is not only affected by the complex institution it accommodates, but external agents as well, such as banking law and changes in technology. Alterations in its environment, whether internal or external to the bank, have implications to its form.

The drive-through teller serves as an example showing how the totality of space adapted to accommodate changes in consumer habits due to the automobile. This addition, a bank's reaction to external factors, had further repercussions to its operations and to the habits of its customers.[69] Although, the totality of space changed due to shifts in its environment, it reciprocated adjustments as well. The bank had to adopt new procedures. Customers developed newer habits in regard to their car, and more importantly, in their expectations of service from a bank. These expectations coupled with further advancements in technology lead to ATMs, a new kind of space for customer service that can serve as an example in which the impetus for change is both internal and external to the bank. Externally, advancement of technology offers new possibilities of space and connection. Internally, banks shifted policy out of a desire to reduce resources, including both customer service space and employees.

ATMs illustrate another characteristic in the relationship between the totality of space and the system to which it belongs. Contrary to expectations, the introduction of ATMs resulted in an unexpected organization of space, with increase in numbers of both ATMs and branches. The experimental introduction of ATMs tested customer's affinity for this new service space. Based on customers' positive response, banks expanded their ATM networks and services offered by their use, with the expressed goal of reducing the number of branches by minimizing customers' dependency on the physical space of branches. In fact, demand for both ATMs and branches increased, resulting in a reassessment of the latter to serve as advertisement as well.[70] Within unit banking practices in the first decades of the 20th century, it was well understood that the bank window functioned as advertisement, as with all commercial businesses. As large branch banks and banking holding companies became dependent on numerous other forms of advertising, they did not realize that this practice had updated itself to include the entire branch as advertisement.

This example demonstrates a moment of self-organization. According to Paul Cilliers, "the capacity for self-organisation is a property of complex systems which enables them to develop or change internal structure spontaneously and adoptively in order to cope with, or manipulate, their environment."[71] In this instance, the bank is attempting to predict the actions/needs of its customers, in order to predict changes to its totality of space. As each customer acts in his or her best interest based on the same set of variables, an aggregate behavior of the group, which reflects the best organization between the system of customers and their environment, in this case, the bank, emerges. Taken as a group, customers exhibit shared behaviors even though they never directly interact with each other. This process is known as self-organization. Such moments within the system of a bank, or within any of its sub-systems, often translate into specific organization within its space. In this case, the introduction of ATMs organized the totality of space by increasing both the number of ATMs and branches.

The process of self-organization exemplified by the introduction of the ATM is one that has appeared numerous times in bank history, as with the introduction of online banking, an immaterial space that again did not reduce the need for traditional bank spaces. In fact, their numbers kept rising, instead of falling or even staying stable.[72] Yet again, where changes in banking procedures predicted specific alterations within the organization of the totality of space, the totality continued to evolve in a different direction. Online banking was an attempt to make customers independent of physical space by interposing an immaterial space between them and their bank. A similar attempt was made in the early decades of the 20th century through banking by mail, an effort by numerous mostly unit banks to extend their customer base by expanding their reach through available communication technologies to loci to which their physical presence was prohibited by banking law.[73] With banking by mail, finally abandoned, a bank's space extended to include the immaterial space of connection to its customers allowed by the technology of mail. When this practice was abandoned by the early unit banks, the form of the bank's space returned to a form similar to its previous one. The external parameter of available technology, on which a bank's self-organization is dependent, led to different final outcomes in the above two cases.The change in the speed of connection created an immediacy that allowed a connection to emerge as a virtual space, whereas in banking by mail the totality of space expanded through a space of connection, with online banking the totality of space expanded to include a virtual instance of the bank's space experienced by its customers.

At any point of time, the current form of the totality of space is sensitive to its history as extended by and integrated with the history of the bank it supports. The path of evolution followed by the totality of space through critical management decisions in relation to restrictions or stimulations imposed by external factors determines possible future outcomes. This does not suggest that history determines future results. It is simply one of many variables that influences future forms. The form of totality of space is guided by a synthesis of its history and by factors both external and internal. As Paul Cillier notes, however,

> . . . the notion of history, however, should be used with care when analyzing a system. It should not be seen as providing a key to final descriptions of a system. This is because the history of a system is not present in the system in such a way that it can be reconstructed. The 'effects' of the history of the system are important, but the history itself is continuously transformed through self-organizing processes in the system - only the traces of history remain, distributed through the system.[74]

History here does not refer to the classical notion of history, a sequential listing of events, but to the effects or traces left within a system due to its previous activities. These effects then serve as parameters that influence future evolution of the system. An understanding of the totality of space can, therefore, arise from the study of both major internal and external factors along with echos of its history. This can serve as basis to predict possible future evolutions of a bank's totality of space, but it should never be assumed to offer a reliable image of the future.

Bank mergers, such as the merger between JPMorgan Chase and Bank One, illustrate history's effect.[75] Both banks were already organizations that evolved through a series of mergers and acquisitions. Where Chase initially had 529 branches and 1730 ATMs, through its merger with Bank One, the resultant company extended its spatial presence to 17 states with 2508 branches and the second largest ATM network with 6650 ATMs. The merger had further implications for the totality of space through the reorganization of offices. The resultant company was headquartered in New York, where Chase had its headquarters, while Chicago, which originally headquartered Bank One, was to headquarter the merged company's retail and middle market business. The new bank also had to define its brand, where Chase invoked stability and tradition, Bank One brought innovation to mind. The resultant brand, under the name Chase, attempted to capitalize on both these aspects. Although, the space emerged was, to a great decree, dependent on Bank One's space, the new bank's spatial character was greatly influenced by the Chase brand developed by JPMorgan Chase. The new bank and its space incorporated characteristics found in both its predecessors. Its eventual totality of space, while dependent on both prior histories established itself as an entity, which continued a path of evolution, constructing its own history.

In addition, the totality of space impacts a bank's identity, which should not be confused with its brand. Banks have already demonstrated an awareness of the need to control space as a means of directing their identity, as seen through the design of buildings that house their headquarters or central offices.

Even though this practice is dependent on space and it affects the identity revealed by a bank's totality of space, this practice is actually an extension of their branding. While banks attempt to rigidly control a customer's perception through advertisements and design, mostly of service spaces, identity refers to the overall impression of a bank as it emerges from elements both within and outside of a bank's control. The impression created by a bank's totality of space is not limited to the effects of branding. Bank decisions that affect changes in any part of their space as a whole, such as the location of its headquarters or other operational offices and the types of locations in which branches operate, each have an impact on the resulting identity. For instance headquarters in New York, NY might imply a level of urban sophistication, while headquarters in Charlotte, NC might imply a connectedness to common people. Along these lines, the merger between NationsBank and BankAmerica offers an indicative example. The resulting company named Bank of America chose to keep NationsBank headquarters in Charlotte, NC over BankAmerica's headquarters in San Francisco, CA, opting for a city that was experiencing new, yet reliable, economic growth in the southeast over a more established economic center of the west coast.

A bank's "personality" can also be seen through the distribution of its branches. Heavy concentration in urban areas results in a very different identity than numerous branches spread throughout rural communities. Trends found in the distribution of a bank's buildings, such as the economic situation, and local industry, affect the bank's identity. In these cases, identity is not a result of stereotypes that belong to each location, but of the activities a bank engages in due to its relationship to specific types of places. For instance, the decision to expand into rural areas necessitates new practices and products demanded by the new customer base.

Furthermore, identity is affected by the nature of technology a bank employs, specifically, by the manner it adopts new technology into its operations. Recently, Chase Bank was the first US bank to introduce banking by SMS.[76] If this endeavor is successful, it suggests that Chase has a strong young customer base. As the first adopter of this technology, its identity becomes one connected with the population that frequently communicates through text messages. In general, technology adopted by a bank relates to its totality of space through the connections it allows. In this case, SMS banking allows quick and temporary access to a virtual extension of the bank's space.

The effects on identity of each change in a bank's space again exhibits the non-linear nature of cause and effect, requiring understanding of the whole in order to truly understand the effects. Identity as seen through the totality of space does not suggest that it can be rigidly controlled. It merely suggests that every decision that affects space also has repercussions on a bank's identity.

8. Three Factors in the Formation of the Totality of Space

As a subsystem within the system defined by the bank itself, the totality of space is most directly influenced by three factors: banking law, management and operational procedures, and technology. These factors act on this space on at least one of two levels: actualization and control. Actualization refers to any factor that helps define the "physical" manifestation of the totality of space, determining the actual form the space adopts, whereas control refers to any factor that imposes limits and directs spatial outcomes. At any given point, these three factors may act on the totality of space to actualize, control, or do both. Each has its own degree of dynamism, which is critical in the formation of an organization's space as a whole. Dynamism refers to each factor's internal propensity for change, measuring the factor's internal rate of change and therefore its potential to affect change within the totality of space.

The space that accommodates a bank's operation as a whole evolves based on desires as they are delineated through its specific actions on and reactions to phenomena that in general are heterogeneous. Each bank is organized through its procedures of operation and administration that form the basis of its function. These procedures are representative of the economic and managerial streams of thought applied by the administration. These streams of thoughts are external to the field of banking, and their advancement and application does not specifically concern financial institutions, but the organization, operation, and activities of corporations in general. The adoption of these principles in a bank's operation culminates in the formation of a central, internal factor, the procedures of operation and management. Procedures are specific in nature and representative of each corporation's personality. As such, they constitute an internal factor, acting as a medium of actualization for a bank's space as a whole. The operational procedures generated by the unique implementation of economic streams of thought driven by the personality of a bank's administration have visible results in the actualization of the bank's totality of space. The organization of a financial institution is obliged to operate at the level of control in order to maintain coherence and consistency. Ultimately, an organization's procedures are an internal factor that both actualizes and controls its totality of space. Its specific form and its effects are dependent on the bank's character, actions, and strategy.

The different strategies of Bank of Italy and Wells Fargo at the beginning of the century illustrate this point. Where the former pursued an aggressive philosophy of acquisition, the latter pursued a more conservative policy of branching and correspondence.[77] Although, Wells Fargo expanded its branching in established urban centers, it feared that acquiring rural banks in remote areas would upset its correspondence relationships. Bank of Italy, which as a new bank, did not have established correspondences with other banks, found that it was far easier to purchase rural banks than to negotiate correspondence. It chose a policy of spreading into bankless small towns and military bases.[78] The managerial philosophy of each bank lead to different organizational structures and procedures, which in turn resulted in different forms for each totality of space. The result at that time was one system in which multiple independent banks related as equals and one in which the system is formed with a central bank to which all the others answered.

Beyond the internal factor of operational procedures, two external ones, banking law and technology, place frames on the development of a bank's totality of space. Banking law acts primarily at the level of control, creating frames that vary widely from country to country, and in the case of the United States, from state to state. These frames are critical in the formation of the spatiality of each bank as banking law systematically intervenes. In the case of the United States, its effect enforces parameters on the formation of the geographic growth of space (whether and under what conditions a bank can establish branches in a city, county, state, or national level), on the type of physical space (whether a bank beyond its headquarters can establish branches, additional offices, etc.), and on the synthesis of spaces, geographic and physical, due to bank mergers and acquisitions, while also placing limitations on the nature of connections between spaces. Additionally, the evolution of law usually reacts to and follows the evolution of management's organization of banks. As such, it is the most conservative of the three factors.

In the United States, banking law was initially structured as a seemingly complete set, whose parts were then updated at irregular, sporadic moments in time, often long after the original impetus for change. It distinguishes itself from the other two factors, as the only one that exists exclusively for banking. Though, it may be external to a specific bank, it functions as an internal factor to the field. The effect of banking law in the United States on a bank's totality of space is apparent in the initial choices made concerning the type of a bank. Banking law defines both the types of banks available (e.g., trust company, commercial bank, and saving bank) and the specific functions allowed for each type, translating to corresponding adjustments in the bank's totality of space. Even at the beginning of the century, banking law dictated that the choice of a savings bank, whose business is confined to accepting deposits with a 2–4% return, translates to a space that could only include a customer service area for deposits and withdrawals, and not a loans' area. Commercial Banks and Loan and Trust Companies, whose businesses are far broader, could have space in their branches dedicated to loans as well as other services.[79] At another level, banking law affects the totality of space in a bank's choice of banking system; whether it is a unit or multiple bank, and in the case of multiple banking, whether it is chain, branch, or group banking. The case of Bank of Italy is indicative here. In order to circumvent laws that prevented chain banking, it invented a new form of multiple banking, group banking. The resultant totality of space was one, which actually exhibited greater unity through management than would have occurred had chain banking been allowed.

The third factor, technology, is externally imposed on bank operating at the level of actualizing the totality of space. Through its various versions (i.e., information technology, communication technology, construction technology, etc.) it supports the materiality of the totality of space. It is produced and advanced externally of the financial institution. A bank can choose which technologies it adopts, how it implements them, and in some cases, how it develops them. An indicative example is that of Bank of America's adoption of ERMA, described earlier in this chapter. New developments in technology allowed it to develop a new application, which had a profound effect on its procedures and connections between its locations. Technology's advancement in most cases imposes the need for frequent updates. Bank of America illustrates this point when it decommissioned ERMA in favor of two IBM mainframes, which of course included shifts in procedures and totality of space. Therefore, the development of the totality of space is in a dependent relation with the advancement of technology.

Technology exhibits a much higher degree of dynamism than either banking law or operational procedures. This is mostly due to the multiplicity of the sources that contribute to its advancement and to the breadth of its applications in a wide variety of fields. Of course, advancement in technology can trigger changes in either of the other factors, as it may suggest new directions for management and operation procedures or necessitate new laws based on novel abilities. For instance, imaging technology coupled with network communication technology made it possible to create and distribute images of bank documents such that any location within a bank could process a document. When this technology was applied at a global level, a bank's totality of space could seamlessly operate 24 hours a day, because work left by an office in one part of the world when it closed could be continued in another part of the world by offices that were just opening. This relationship can also occur in the opposite direction. Shifts in either of the two other factors can create shifts in the development of new technology. For instance, ATMs are a space and a technology that was developed at the behest of banks based on their interest in automated 24-hour customer service both for reasons of better service and cost reduction.

These three factors, management and operational procedures, banking law, and technology are pivotal in the systematic formation of a bank's totality of space through their actions in its actualization and control. As a group, they form the administration's tools and limitations in designing the space in which a bank operates. Additionally, the degree of dynamism they each exhibit is of great importance in the formation of every spatial result in a bank's totality of space, which shows its own degree of dynamism, composed by the degrees of dynamism exhibited by each factor. A study of a bank's totality of space should occur through its relationship to these three factors. Furthermore, its potential future forms are tied to the three factors and their evolution.

9. Conclusions – A Possible Architectural Response

Although financial institutions imply and use characteristics of the totality of space, they lack the proper training in spatial thought necessary to effectively describe and therefore fully understand the properties of this "novel" space, which has for sometime sheltered their activities. Establishing these characteristics allows architects, together with their clients, to understand this space, manage it correctly, and direct changes designing within it. A macroscale level of understanding in space design can drive to better decisions at the microscale level. This suggests that a new, essential step in architectural design that needs to be developed has already emerged. Although, currently, the focus is still located on the design of the microscale found at the level of the physical entity, the building, it has become necessary for a design process of the totality of space to take precedence. This is a process that accounts for a macroscale level of design that manages the entire space of a bank. This kind of design should account for the three basic factors, banking law, management and operational procedures, and technology.

The spatial mapping or diagramming of the development of the specific relationships that each of these three factors form with an individual bank is a vital component of macroscale level analysis, in that it allows the understanding of the current as well as past formations of its totality of space. The history of these three factors, as they interweave to configure the formation of this space, reveals potential directions for its future evolution. These factors, ultimately, play a dual role in the macroscale level of design. They help in understanding the form of a bank's space as a whole, and they form a set of tools that can be used in directing this space toward specific potentials. It is the synthesis formed by maps and diagrams of the three factors in relationship to a specific bank that renders the complete image and potential futures of its totality of space. Therefore, the diagraming or mapping of their historic development, in relation always to a specific company, can comprise an important architectural tool in the direction of macrolevel design.

As a first step in the analysis of a bank's totality of space architects should deal with the connections that define the entirety of space, which accommodates function and identity. These connections are defined by operational procedures, as well as the two external factors, banking law and technology. In fact, procedures, as they adapt to changes in banking law and technology, ultimately define the bank itself. Although, within a specific field operational procedures are nearly identical, each company implements them uniquely according to its needs, philosophy, and the relationship it chooses to build with each external factors. Space is developed, while a financial institution organizes its body of operation, reacts to banking law, and internalizes technology, thus dictating procedures necessary to divide, organize, connect, allocate and direct its parts and its resultant form. The result is unpredictable, but according to Rifkin[80] and Windschuttle[81] it is understandable. As this process extends into space, understanding its path together with its effects allows architects to consciously assist corporations to manage existing space and direct changes to the totality. While management decisions direct a company's future development, the right procedures provide a solid foundation for steady transitions keeping a bank's whole coherent and operational. These procedures, in no case, help predict a company's exact future form. They can only guide future developments. In the field of architecture, this observation translates to exploration of space procedures that can guide the development of physical and virtual structures and connections for financial institutions.

Macroscale level design should address two directions, which are in fact intertwined within the totality of space. On the one hand, it deals with the function of the totality of space that supports the operations of a bank. Here, space has to be managed as an internal process, integral to a corporation's operations. On the other hand, it must account for the identity of a company as an external interaction emerging from totality. After all, identity does not reside within the separated monads of the company's space. Space design should include procedures that guide future forms, since it can no longer limit itself to the design of individual structures.

The totality of space inherits self-organizing characteristics from its relationship with the institution it hosts. This fact presents a conflict, how is it possible to design, when it is impossible to predict the final appearance of any? The answer lies in the premise that it is possible to design in order to direct and accommodate future self-organization, according to the needs of a constantly evolving institution. It therefore suggests the necessity to develop guides that can be used to understand, manage, and direct the design of these new types of spaces. The architect's role is no longer limited to the construction of new buildings, nor does it expand to include the design of the corporation. This is an issue for a company's governing body. Architecture's new task is the analysis of a company's existing spatial utility (the creation of its space maps and diagrams) in order to direct the design of additions or subtractions, as well as the proper translation into space of the company's character. The architect's role in the management of institution's existing space is crucial. Moreover, his contribution in design in order to direct future unpredictable changes in space is imperative. As already stated, this sort of design is currently undertaken unconsciously by corporations themselves, but the inclusion of architects in this process will prompt more conscious and therefore more successful results.

Notes

1. Parts of this chapter have been previously published in the conference proceedings of 3rd IET International Conference on Intelligent Environments, (2007), in a paper entitled Corporate Space as a Totality: Technology's Effect on Spatial Thought.

2. The Shaw Banking Series, *Building Equipment and Supplies*, xv.

3. Examples of such books include, Fiske, *The Modern Bank: A Description of its Functions and Methods and a Brief Account of the Development and Present Systems of Banking* and Corns, *The Practical Operations and Management of a Bank*. Many of these books go so far as to include floor plans. Even manufacturing companies, which produce banking equipment, tried to express their views on the subject, by publishing books of their own concerning the architecture of such a building. U.S. Bank Note Company, *How to Organize and Equip a Modern Bank*, is an example of one such book.

4. MacGregor, *Bank Advertising Plans*, 88.

5. Belfoure, *Monuments to Money: the architecture of American banks*, 129–131.

6. Belfoure.

7. Belfoure, 129–131.

8. Hopkins, *The Fundamentals of Good Bank Building*, 1.

9. Belfoure, 132.

10. Morehouse, *Bank Window Advertising*, 5.

11. Morehouse, 7.

12. Shaw, 13.

13. Shaw, 31.

14. Shaw, 31.

15. Belfoure, 245–246.

16. Belfoure, 249.

17. Belfoure, 280.

18. Belfoure, 294–296.

19. Belfoure, 291–292.

20. See Cartinhour, *Branch, Group and Chain Banking*, 104, and Southworth, *Branch Banking in The United States*, 58. Both suggest that the most progressive banks of that era were pursuing some form of multiple banking.

21. Southworth, 5.

22. Mason, *The Transformation of Commercial Banking in the United States*, 24.

23. See Mason, 24 and Southworth, 10–16. The National Bank Act of 1863 contained no language directly referring to branch banks and in fact contained language that created ambiguity regarding branches for national banks. The 1864 revision still did not refer to branches, but corrected this ambiguity such that branches were not permitted.

24. Willit, *Chain, Group and Branch Banking*, 15.

25. See Mason, 24. For a detailed analysis of branch banking in selected states, see Southworth. For a state by state analysis of chain and group banking, see The Economic Policy Commission of American Bankers Association, *A Study of Group and Chain Banking*.

26. Willit, 15.

27. Willit, 18–19. Statewide branch banking was allowed within 9 states. Limited branch banking was permitted in 10 states. Branch banking was prohibited in 22 states, 8 of which allowed pre-existing branch banks to continue operation of their branches. The remaining 7 states left branch banking unregulated.

28. See Collins, *The Branch Banking Question*, pp. 53–56. In some cases, state banking law interferes with the physicality of the types of space a bank operates, by legislating the specific operations a bank may include in its additional spaces. For instance, additional offices or limited branches are restricted to the services of deposits and withdrawals, a set of functions similar to the first ATMs.

29. Willit, 18–19.

30. House Committee on Banking and Currency, *Branch, Chain, and Group Banking: Hearings before the Committee on Banking and Currency*. 71st Congress, 2nd Session, 26.

31. See Collins, 25–26, and Cartinhour. According to the Bank of America 1964 Annual Report, between its establishment in 1904 and 1926 Bank of Italy grew from one office to 98 branches. Also see James and James, *Biography of a Bank: The Story of Bank of America N.T. & S.A.*, for an analysis of the manner in which L.M. Gianini, the founder of Bank of Italy, additionally founded three other separate legal banking entities, in response to legal restrictions. In 1927, changes in banking law allowed these four entities to legally merge into one under the name Bank of Italy.

32. Southworth, 96–98.

33. Southworth, 94.

34. Mason, 26.

35. Mason, 26.

36. See Rousakis, *Commercial Banking in an Era of Deregulation*, 51–55. By organizing themselves as holding companies, banks could effectively cross state lines. Some of the first banking holding companies included companies that offered non-banking services.

37. Fewer than 50 multibank holding companies were established during the 20 years prior to 1956. The decade following the Bank Holding Act of 1956 experienced the founding of 31 multibank holding companies, and the 5 years following the enactment of the Bank Merger Act of 1966 included 111 such foundings. See Mason, 29–33.

38. The Bank Holding Act of 1956 was primarily interested in holding companies that operated multiple banks. Such institutions were only allowed to own interest in companies closely tied to banking and financial services. See Prochnow, *The One-Bank Holding Company*, 20–21.

39. Mason, 31.

40. The first financial institutions to gain the right to cross state lines were thrifts in 1992, 5 years prior to the rest of other types of banks. This was due to changes in regulations enacted by the Office of Thrift Supervision. See Rousakis, 45.

41. This conservativeness continued in the '1990s. Banks still adopted emerging technology through tentative steps, but now they could seek the advice of business technology experts. See Slater, Choosing the Right (IT) Path, A-12.

42. See American Technical Society, *Cyclopedia of Commerce, Accountancy, Business Administration, Volume 1*, 159–162, which mentions that some banks found phones useful for interdepartmental communication, but that much business was better conducted through written documents as opposed to verbal communication. Written documents immediately served as a record of a transaction, protecting from miscommunication and misunderstandings.

43. For further information on the development of banking forms during this era, see American Technical Society, 288–289; The Shaw Banking Series, 170–193; United States Bank Note Company, 43–48; and especially Law Journal Year Book, *Banking Forms: A Practical and Up-To-Date Set of Banking Forms, Arranged in Groups Covering the Various Departments in a Commercial Banking Institution, Including Trust Company Forms and Savings Bank Forms*.

44. For a full description of the development of filing systems during this era, see American Technical Society, 257–261, The Shaw Banking Series, 144–147 and United States Bank Note Company, 49–53.

45. American Bankers Association, *Commercial Bank Management Booklet No.14*, 4. Emphasis as in original.

46. American Bankers Association, 5.

47. American Bankers Association, 5.

48. American Technical Society, 135.

49. James and James, 110–118.

50. Bank of America, *Annual Report, 1962*, 19.

51. Bank of America, *Annual Report, 1960*, 19.

52. It should be noted that distance was still an issue. Even as ERMA was being developed, Bank of America was testing semi-automatic bookkeeping machines to support remote branches with accounting problems, predicting that ERMA would not be able to serve all of its branches. See Bank of America, *Annual Report, 1958*, 18–19.

53. Bank of America *Annual Report, 1966*, 7.

54. Bank of America, 1966, 7.

55. Credit cards were promoted as a better way for customers to manage their credit, concentrating their debts into one bill, and as a service to merchants who instead of offering credits could use their capital to expand their business. See Bank of America, *Annual Report, 1959*, 19.

56. Kratz (1975): Walter Wriston automates the teller.

57. See Kratz. Also, Vesala, Banking Industry Performance in Europe: Trends and Issues, 125–126 includes a discussion on the expansion of ATMs as a means of attracting new depositors.

58. O'Heney, The Wired World of Trade Finance, A-20 - A-22.A-20–A-22.

59. See O' Heney, A-20–A-22. For a discussion of Imaging Technology after it had been widely adopted by the banking industry, see Ooley, Checking It Twice: Check Imaging System Offers Greater Flexibility and Efficiency.

60. Security First Network Bank on October 18, 1995 became not only the first bank to offer Internet Banking, but the first virtual bank as well. For a detailed analysis see Humphreys, Banking on the Web: Security First Network Bank and the Development of Virtual Financial Institutions, and Christopher, Recent Developments Affecting Depository Institutions.

61. Rockefeller, *Creative Management in Banking*, vii.

62. Saskia Sassen, (1994), *Cities in a World Economy*, London, BGR: Pine Forge Press.

63. Sassen, The Global City: Strategic Site/New Frontier

64. Cilliers, *Complexity and Postmodernism : Understanding Complex Systems* 4.

65. Cilliers, viii.

66. Bank of America, 1964.

67. James and James, 194.

68. James and James, 310.

69. Bank of America, 1958, 18.

70. Thomke and Nimgade, Bank of America (A), 4.

71. Cilliers, 90.

72. Thomke and Nimgade, 4.

73. Ingraham, *Bank Advertising or Methods of Procuring New Accounts and Other Profitable Business*, 79.

74. Cilliers, 108.

75. Bank One *Annual Report, 2003*; JPMorgan Chase, *JPMorgan Chase Annual Report, 2003*; and JPMorgan Chase, *JPMorgan Chase Annual Report, 2004*.

76. JPMorgan Chase, TXT MSGS MAKE BNKG EZ: Chase Mobile Gives Consumers Real-time Account Info.

77. Southworth, 58.

78. Southworth, 58.

79. Coffin, *The A B C of Banks and Banking*, 21–28.

80. Rifkin, *The Age of Access*.

81. Windschuttle, *The Killing of History*.

References

American Bankers Association. (1934). *Commercial Bank Management Booklet No. 14*. New York: Clearinghouse Section, American Bankers Association.

American Technical Society. (1912). *Cyclopedia of Commerce, Accountancy, Business Administration, Volume 1*. Chicago, IL: American Technical Society.

Bank of America. (1958). *Bank of America Annual Report.*

Bank of America. (1959). *Bank of America Annual Report.*

Bank of America. (1960). *Bank of America Annual Report.*

Bank of America. (1962). *Bank of America Annual Report.*

Bank of America. (1964). *Bank of America Annual Report.*

Bank of America. (1966). *Bank of America Annual Report.*

Bank One. (2003). *Bank One Annual Report.*

Belfoure, C. (2005). *Monuments to Money: the Architecture of American banks*. Jefferson, NC: McFaland & Company, Inc.

Cartinhour, G. T. (1931). *Branch, Group and Chain Banking*. New York: The MacMillan Company. Republished in Bruchey, Stuart W. and Carosso, Vincent P. (eds.). (1980). *The Rise of Commercial Banking*, New York: Arno Press.

Christopher, B. B. (1996, February). Recent Developments Affecting Depository Institutions. *FDIC Banking Review, 8*(3). Article III. Retrieved November 17, 2007, from http://www.fdic.gov/bank/analytical/banking/1995summ/art3full.html.

Cilliers, P. (1998). *Complexity and Postmodernism : Understanding Complex Systems*. London, GBR: Routledge.

Coffin, G. M. (1903). *The A B C of Banks and Banking*. New York: S.A. Nelson.

Collins, C. W. (1926). *The Branch Banking Question*. New York: The MacMillan Co.

Corns, M. C. (1968). *The Practical Operations and Management of a Bank* (2nd ed.). Boston, MA: Bankers Publishing Company.

The Economic Policy Commission of American Bankers Association. (1929). *A Study of Group and Chain Banking*. New York: American Bankers Association.

Fiske, A. K. (1904). *The Modern Bank: A Description of Its Functions and Methods and a Brief Account of the Development and Present Systems of Banking*. New York: D. Appleton & Company.

Hopkins, A. (1929). *The Fundamentals of Good Bank Building*. New York: The Bankers Publishing Company.

House Committee on Banking and Currency (1930). *Branch, Chain, and Group Banking: Hearings Before the Committee on Banking and Currency.* 71st Congress, 2nd Session.

Humphreys, K. (Security First Network Bank). (1997). Banking on the Web: Security First Network Bank and the Development of Virtual Financial Institutions. In Cronin M. (ed), *Banking and Finance on the Internet.* (pp. 75–106). New York: Wiley, John & Sons Inc.

Ingraham, A.M. (1911). *Bank Advertising or Methods of Procuring New Accounts and Other Profitable Business.* Cleveland, OH: A.M. Ingraham, Financial Advertsing.

James, M. and James, B. R. (1954). *Biography of a Bank: The Story of Bank of America N.T. & S.A.*. New York: Harper & Brothers.

JPMorgan Chase. (2003). *JPMorgan Chase Annual Report.*

JPMorgan Chase. (2004). *JPMorgan Chase Annual Report.*

JPMorgan Chase. (2007, September 19). TXT MSGS MAKE BNKG EZ: Chase Mobile Gives Consumers Real-Time Account Info. Chicago. Retrieved October 19, 2007, from http://investor.shareholder.com/jpmorganchase/press/releasedetail.cfm?ReleaseID=264788&ReleaseType=Current.

Kratz, E. F. (2005, June 27). 1975: Walter Wriston Automates the Teller. *Fortune, 151*(13).

Law Journal Year Book. (1917). *Banking Forms: A Practical and Up-To-Date Set of Banking Forms, Arranged in Groups Covering the Various Departments in a Commercial Banking Institution, Including Trust Company Forms and Savings Bank Forms.* New York: Banking Law Journal Co.

MacGregor, T. D. (1913). *Bank Advertising Plans*, New York: The Bankers Publishing Co.

Mason, J. E. (1997). *The Transformation of Commercial Banking in the United States.* New York: Garland Publishing Inc.

Morehouse, R. W. (1919). *Bank Window Advertising.* New York: The Bankers Publishing Company.

O' Heney, S. (1993, February). The Wired World of Trade Finance, *Bank Management, LXIX*(2), A-20–A-22.

Ooley, D. (1998). Checking It Twice: Check Imaging System Offers Greater Flexibility and Efficiency. In Keyes, J. (ed), *Banking Technology Handbook.* (7-1–7-5). Boca Raton, FL: CRC Press LLC.

Prochnow, H. V. (1969). *The One-Bank Holding Company.* Chicago, IL: Rand McNally & Company.

Rifkin, J. (2000). *The Age of Access.* New York: Tarcher/Putnam.

Rockefeller, D. (1964). *Creative Management in Banking.* New York: McGraw-Hill Book Co.

Rousakis, E. N. (1997). *Commercial Banking in an Era of Deregulation* (3rd ed.). Westport, CT: Praeger Publishers.

Sassen, S. (1994). *Cities in a World Economy*. London, BGR: Pine Forge Press.

Sassen, S. (2001). The Global City: Strategic Site/New Frontier. Retrieved March 2, 2007, from http://www.india-seminar.com/2001/503/503

The Shaw Banking Series, (1919). *Building Equipment and Supplies*. New York: A.W. Shaw Company.

Slater, R. B. (1993, February). Choosing the Right (IT) Path. *Bank Management, LXIX*(2).

Southworth, S. D. (1928). *Branch Banking in The United States*. New York: McGraw-Hill Book Company, Inc.

Thomke, S. and Nimgade, A. (2002). *Bank of America* (A). Boston, MA: Harvard Business School Publishing.

U.S. Bank Note Company. (1913). *How to Organize and Equip a Modern Bank*. Indianapolis, IN: U.S. Bank Note Company.

Vesala, J. (1995). Banking Industry Performance in Europe: Trends and Issues. In *OECD Documents, The New Financial Landscape: Forces Shaping the Revolution in Banking, Risk Management and Capital Markets*, (97–165). Paris: OECD.

Willit, V. (1930). *Chain, Group and Branch Banking*. New York: The H.W. Wilson Company.

Windschuttle, K. (1996). *The Killing of History*. San Francisco, CA: Encounter Books.

Index

GPSR Compliance

The European Union's (EU) General Product Safety Regulation (GPSR) is a set of rules that requires consumer products to be safe and our obligations to ensure this.

If you have any concerns about our products, you can contact us on ProductSafety@springernature.com

In case Publisher is established outside the EU, the EU authorized representative is:

Springer Nature Customer Service Center GmbH
Europaplatz 3
69115 Heidelberg, Germany

Batch number: 10755284

Printed by Printforce, the Netherlands